Indiana Bigfoot reports and the book of Kochi in the same book is a conflicting work of two worlds, one of the flesh and blood and one of a world of seamless messages and sightings from one man and several of Indiana residents.

It is the author's job to report these facts and incidents as they come in, but there is a problem there as well, it takes several years in order to collect enough reports through the ones that are finally reported through the thousands of reports that never get reported, because of ridicule and nasty remarks from close minded people, and through religious dogma, we are taught that things are monsters or the work of the devil, which is not true at all, the author is a Christian that does not go to church because of the different ideals of Christ, to the author, one in Christ is everything, the blood, salvation through that blood, after ten years of worship and love for the lord, the author , realizes that there is more to this world that meets the eyes or heart. So to that end means Steve has undertaken the truth as he sees it, and has reported it in this book, I have taken it upon myself to copy and paste some of the reports and those will be put first in the book as they came in, and will try to make them in alphabetic order of the counties, to make it easier to read. and then the authors story last, that way, the reader can read about reports and incidents that pertain to the flesh and blood creature, and then the story last, so the reader will know the way the forest dweller lives among humans, some of the reports will be in smaller wording throughout the book, because of the copying and pasting,

S it will look a little unprofessional, just so you understand how to read the book at times, so let's begin,

Morgan County, Indiana 1991

Morgan Monroe State Forest, Martinsville, Indiana

Hidden in the wilds of Morgan Monroe State Forest is a body of water called Lost Lake. Even the locals have trouble finding it , there is no road to it.

A group of three men coon hunted this area regularly and had for over 25 years. They drove in so far and hiked the final half mile to the lake. Their coon dogs seemed to get very excited as they approached the hidden lake and the men figured there has to be coons everywhere.

~ 4 ~

One of the men that had one of the dogs pulling very hard on its lead kept thinking that something in the forest was following him.

Finally the men decided it was time to let the dogs go before they went crazy. The moment the dogs were released there was a loud crack that came out of the forest right behind them.

When they turned around to see what had made the noise they were about fifteen feet away from what had to be the largest man they had ever seen. Covered with hair the man was about nine feet tall and was staring them down and it was not at all afraid of the three men. The men stood motionless and watched as it simply turned around and walked in to the darkness of the forest. The men finally got their dogs called in and very quickly hiked back to their trucks.

They never hunted this area again and said they would never return to Lost Lake day or night.

++++++++++++++++++++++++++
++++++++++++++++++++++++++

Morgan County, Martinsville , Indiana

Morgan Monroe State Forest 1984

A college student attending Indiana University in Bloomington, Indiana was living in a small house that's property line bordered Morgan- Monroe state forest. Since he had moved in, he spent little time there during the day and basically it was a place to sleep. But sleep was getting difficult because of the sounds that would awaken him that came from the forest. The high piercing screams would usually send him out of his bed and running for the nearest light.

Friends that would visit on the weekends said that they saw something running across his driveway behind their cars in their rearview mirrors. Rocks were being thrown on his roof at night and the constant pelting sound made it impossible to sleep.

Many times he could hear the sound of something walking around his home accompanied by the sound of loud breathing. He would walk outside look into the darkness but he saw nothing. So he decided to order a pair of night vision glasses from army surplus.

Every time he would try to use them the sounds he was hearing would stop. It was getting obvious whatever he was hearing could see the beam from the night vision glasses.

Then one night he got home very late and he had the night vision glasses in his backseat, for some reason he picked them up turned them on and started looking through them.

The first thing he saw was standing in his driveway about twenty feet away and was staring directly at him. It was at least 7 feet tall, large head, no neck and covered with long dark hair or fur. Its arms hung to its knees at its sides and it stood motionless. What he described as a diamond shape pupil that he could see thru the glasses, shined in the light of the one outside security light. He just stared at it and he did not move, then it turned around took two steps and stepped behind a tree, and it did not move. What he noticed was the color of the hair was darker between its shoulder blades and down its back. The young man could still see the side of it. It seemed to be waiting to see if it was being followed. The man backed up and walked into his house not taking his eyes off of the hairy shoulder that could still be seen from behind the tree.

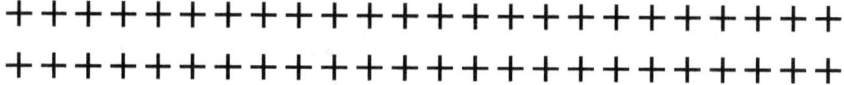

Morgan County, Indiana 1991

Morgan Monroe State Forest, Martinsville, Indiana

Hidden in the wilds of Morgan Monroe State Forest is a body of water called

Lost Lake. Even the locals have trouble finding it , there is no road to it.

A group of three men coon hunted this area regularly and had for over 25 years. They drove in so far and hiked the final half mile to the lake. Their coon dogs seemed to get very excited as they approached the hidden lake and the men figured there has to be coons everywhere.

One of the men that had one of the dogs pulling very hard on its lead kept thinking that something in the forest was following him.

Finally the men decided it was time to let the dogs go before they went crazy. The moment the dogs were released there was a loud crack that came out of the forest right behind them.

When they turned around to see what had made the noise they were about fifteen feet away from what had to be the largest man they had ever seen. Covered with hair the man was about nine feet tall and was staring them down and it was not at all afraid of the three men. The men stood motionless and watched as it simply turned around and walked in to the darkness of the forest. The men finally got their dogs called in and very quickly hiked back to their trucks.

They never hunted this area again an said they would never return to Lost Lake day or night.

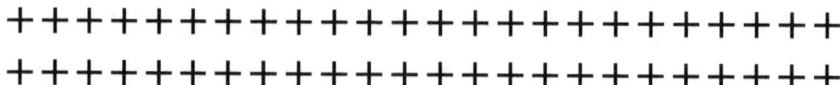

Vigo County Interstate 70 East March 20, 2013

~ 11 ~

Two Young Women Have Late Evening Sighting

DATE: Wednesday, March 20, 2013

TIME: Aprox. 11:30pm

LOCATION: Eastbound Lane - Interstate 70 between Welcome Station/Rest Stop and the Wabash River just west of Terre Haute, IN and east of the Illinois State line

TYPE: Visual - close range

WEATHER: Clear and Cold

AREA/TERRAIN: woods, right by the interstate

WITNESS STATEMENT (taken directly from her website submission) : We were driving down the interstate, and it was very dark. I was on the phone informing my mother of our whereabouts. As soon as I hung up the phone and looked up, the figure was there. It was walking in the opposite direction as we were driving. We were in an SUV but the form was as large, if not taller, than the vehicle. It was very tall, did not appear to have a distinct neck area, and was walking leisurely right beside the road. There were woods all around. It was as large as a bear but walking upright. It resembled an ape-like figure with fur. We believe it could have been searching for dead animals on the side of the road

sounds were heard, the windows were up and the encounter was very brief. We were driving 70 mph, the head area, arms, legs, and a body were clearly outlined... black shadowy with fur all over, almost as tall as my SUV.

When asked what they (both witnesses) did after the sighting: Discussed what we saw and confirmed that we both saw it ... Evaluate what just happened and how amazing it was.

MIKE'S NOTES: I spoke on the phone {with the passenger of the SUV who submitted the sighting report,} on Sunday March 24, 2013 which was only 4 days after the encounter. She was very clear in what transpired back on Wednesday night and still clearly excited about having the experience. Here's my synopsis of both the written report and my conversation with her.

On Wednesday evening March 20, 2013, the 2 witnesses were returning to their homes on the eastern side of Indiana via Interstate 70, having just traveled thru Illinois heading east. They had just crossed the state line and passed Welcome Station / Rest Stop which is about 1 1/2 miles inside the Indiana state line. Somewhere in the next 3 miles BEFORE getting to the Wabash River (Terre Haute is just 1 mile east of the Wabash River) the reporting witness was finishing a phone call with a family member back home to let them know how far from home they still were. As she finished the call and hung up, she looked up to see as they were driving in the right-hand lane of the east-bound lanes of I-70 that there was a figure walking toward them, just off the travel portion of the road, just outside the solid white lane marker. At first (implied as they approached it) SHE thought that there was this bear walking on the side of the road headed west on their right side (passenger side) of the vehicle. But once they passed it (neither speaking yet) she realized it was neither a bear or a person as it was not clothed, was walking on 2 feet and had an ape-like appearance. It didn't appear to attempt to move off the shoulder of the road to avoid them: it just seemed to be patiently walking along the side of the road, in no big hurry After passing, they began confirming between the two of them what each had seen, and with no apparent disagreement, both confirmed to each other a visual description of what had just transpired. The only other thing the witness could report was that the friend who was driving stated she had seen it coming for just a little longer than she (the passenger) had and was actively watching it, as they came upon it while the passenger was finishing her phone call home, looking up in just enough time to see it approach and pass by the vehicle.

While I don't care for the obligitory "the witness seemed credible" type of statements in Bigfoot reports, I have to say that in talking with the submitting witness, that she was very clear and articulate about the event. Her recollections and memory of the event were very clear, and the longer we talked, a sense of excitement was evident in her voice as she relived and talked about her sighting. If in the coming days, she or the driver recall anything else previously unreported, the witness or the driver will be back in contact to add it to the report.

The western side of Indiana, with the Wabash River running down to the Ohio River, is full of reports of sightings and encounters. Several first hand reports I've received haven't made it to the web for differing reasons. Another report from Terre Haute in 1995 will be posting in the Vigo County section very soon. All in all, it's interesting to see over various spans of time, the reports continue to come in from specific regions of the state like the I-70 corridor from Indy over to the Illinois State line.

Enter supporting content here

contact : info@IndianaBigfoot.com

or write:

IndianaBigfoot.com

P.O.Box 420

North Webster, IN 46555

++

Young Woman Has Encounter To Last A Lifetime

I'm presenting this sighting exactly as it was reported to me because of how it was clearly described in a first-hand manner. This is taken directly from my website submission form. While many elements of the encounter are very consistent with other reports, a few answers and descriptions given may surprise you, but are also sometimes experienced by other first-hand witnesses, I've learned over the years to not simply dismiss all reports out-of-hand because some elements of the story seem unusual or don't match my personal experience. With that, please read this report with an open mind and understand that not everything that occurs in life can always be easily explained.

Your Name: xxxx

Address/City/State/Zip: xxxx

E-Mail Address: xxxx

Phone number: xxxx

Your age/gender: xxxx

What other way(s) may we contact you to verify your submission? xxxx

When did this event occur? Date/Time: / How long did it last? - The sighting took place in the fall of 1995

Who (by name) was present when the event occured? xxxx

Where did this occur? Be highly specific. Nearest landmarks? State/local roads? - I lived in a small subivision on the outskits of Terre Haute Indiana. Our home was at the dead end of the subdivision on a hill and behind our house was a very large corn field. There is woods to the left of the field that go on for along ways.

What is the area like? Swamp/Woods/Field/Creek/Open Area/Other - [empty]

What was the weather like at the time? - It was around Halloween time and it was cloudy. The field was very muddy from the recent rain.

Tell YOUR STORY here with as much detail as possible. Include ANYTHING you can remember even if you don't know that it pertains. -   I was board one afternoon because I had been inside for a few days due 2 rain an cold so I decided to go exploring in the corn field behind my house. I had seen a few deer grazing in that field before and I was kinda hopefuL I might get to see some that day.I remember the field had long since been harvested and the ground was really soft and muddy. I am unsure how far out into the field I walked but it was a good ten minute walk to get back.the corn field curved off to the left and there was woods the ran along the length of it. After awhile I spotted two adult deer and a very large buck grazing. I got about 30 feet away from them and I just stood there an watched them. They had seen me Im sure but they didnt seem 2 fazed by my presence I think maybe cause I was a comfortable distance away from them maybe. Then they looked up like they heard something and just took off like they were afraid and I looked at the area they had been looking at and thats when I saw what I then callled a gorilla-man. This creature had a human like face with a flat wide nose and very high broad shoulders. It was walking toward the field andit didnt seem aware of me at first but the moment it looked my way it stopped and stood very still. It was right at the treeline not quite in the woods andbarely in the field so I had a very clear an unobstructed view of it. It was about 20 to 25 feet from me and it was the blackest color I have ever seen only the face seemed a bit lighter.It stood there for at least a full minute and I just stared at it. It stood so still that I began to wonder if I was seeing things an it wasnt a shadow or some bear so I walked closer. When I was about 15feet from it my heart began to race because I could see very clearly it was a very tall hairy animal that was bipedal and had human like features such as its dark eyes and they way it stood was very humanlike. It still wouldn't move though so I reached down an picked up a clump of dirt and tossed it at it. I had no intentions of trying to hit it I just wanted to see if it moved because it was so motionless I was doubting myself what even though I could clearly see it. The dirt clump landed a few feet away from it but it made the creature bend down a little like it was gona dodge it. Then it stood back up fully erect and gave me a very displeased look before it turned and began to walk back into the forest. I was so shocked and curious at the same time so I wasnt really thinking straight and I began yelling hey stop wait as I ran towards the area It was. I stopped right at the beginning of the woods and there was a deep ditch and the creature was already walking up the other side. I yelled at it again and wave and asked it to stop but it just turned and looked at me for a moment and then walked the rest of the way up the ditch and into the woods on the other side. I have no idea why I ran after that thing still to this day I cannot belive I didnt run screaming. It was like the minute it was walking out of sight it just hit me how scary the situation really was and I realized that as huge as the creature was it could of really hurt me badly if it had wanted to. I turned and ran as fast as I could back toward my house and I burst in the front door and started blurting out what had just happened to my mother. While I was still telling her what Id seen the trailer suddenly began to shake violently to the point of knocking a picture off the wall and throwing me off balance and i fell over. When the trailer stopped moving my mom ran outside and I was right behind her because she thought we had just had an earth quake but nobody else was outside and so we ran to the neighbors and asked him if he had just felt anything but him and his wife both said they hadnt felt anything. My mom got really freaked at at that point and about two nights later I was at a friends house and my mom and step dad said the entire trailer began to shake and rock worse than before and my step dad grabed my mother and and made for the front door. They both claim the moment they we outside standing on the wooden deck the trailer stopped moving and they were suddenly almost knocked down by a very sudden and strong blast of wind. The strength of the wind actually knocked the porch swing off the deck and it was made of a heavy metal because it wasnt the type that hung from the roof but it swung from its own base. After this all happened our neighbor claimed about a week later that he had caught a large hairy bipedal creature on his porch looking in his window late at night. the neighbor had a beautiful log cabin looking house and the front had a very large picture window that faced the road pointing the oposite way of the field behind us.

Have you ever experienced anything like this before? If so, explain. - I have never again or before saw anything like I had seen that day but I do have a family member who now lives in that same trailer or modgular and he and his wife both claim to of seen something similar to what I did walking in the field behind their house and once in their own back yard.

Have you had any OTHER odd or strange encounters at THIS location or elsewhere before? - No never.

What did you do when this happened? - I remember telling my mom I had just saw a giant gorilla that walked like a man. It hadnt even dawned on me until a few weeks later that what id seen had fit the discriptions of other big foot sightings.

What else comes to mind "after the fact"? - It seemed to be startled by seeing me just as I was by seeing it. It obviously was interested or curious about me also and it watched me while I waved and shouted hello trying to get a reaction from it but it remained perfectly still.

What is YOUR gut reaction to/about what happened? - I think that day I saw it that was there possiably to hunt deer because there was alot of them that grazed in that field plus the reaction from them when they picked up on it was fear.

What has this event done to change you, if at all? - It has caused me to actually be less critical when I hear other people talk about their own strange encounters

Could you clearly see a head? Face? Eyes? Nose? Mouth? Ear? Other features? Please describe: - It was the blackest color I have ever seen before in my entire life, I cannot explain how black it was. It was much lighter though around the eyes nose and face because it had less hair there. It had completely black eyes and I saw no white in them whatsoever. They we very large for its face and it had a wide flattende looking nose. Its mouth stayed close with very thin lips until the very last time it looked at me before it walked out of sight.It opened its mouth slightly and I could see well enough that its incisor teeth were much longer than the others kinda reminding me of a baboons teeth. its head was shoped alot like a gorillas with a cone type rounded skull. I could see that it was very muscular also kinda reminding me of the upper part of a silver back gorilla also.

If you made "eye contact",, how long did it last? What did you think about it? How did it make you feel? - I made direct eye contact with it for what seemed like 2 minutes before I finaly threw something at it to make it move. Since it had such large pure black eyes my opinion is it is a mainly nocturnal animal.

What did "it" do? What did YOU do? - I think I was in shock but I was just happy I wasnt the only one who experienced something unexplainable at the time. I didnt even tell anyone else because I wasnt sure how to even explain it. It was so unbelievable even to me at the time.

Did you hear any specific sounds? Describe: Did you feel any physical effects on you? Describe: - I did not hear a sound from it the entire time but I still am puzzeled by it sighting I have left out one very big part of this encounter because it is even more unbelievable to me than just the sighting. The last time it turned around when I was begining to follow it into the woods I heard a voice tell me DO NOT FOLLOW ME but even stranger the creatures mouth never moved and im not going to even try and explain it but

I believe it spoke to me directly in my own mind.

Could you clearly make out other features such as arms, legs, frontal regions, hind regions? Unique features? - The cone shaped head reminded me alot of a gorilla I'm sure thats why I refered to it as the gorilla-man for a long time. It had to of stood around 7ft tall and I remember it had very long arms with human like hands and the bottom of its feet did not seem to have hair I notice when it was walking away.

How tall would you estimate it was? What color was it? - It had to of stood at least 7ft because when I got to the same spot where it had been standing I remember its head was even with the brance of the sycamore tree that it stood by. I was 5ft at the time im now

5'2 and I raise my arm up to see if I could touch the branch at its head level and I still could not. It is even harder for me to explain to you just how black it was. I can only say that I have never seen anything that black before or since and had it been night Im positive it would have been completely unvisable.

How did it walk or run? Describe: - [empty]

Did it pickup or move anything? Was it carrying anything? - it was moving through the branches toward the open field until it saw me and stopped

Did you notice any odd, bad, or strange smells during the encounter? Describe: - No I never smelled anything unusual

Anything else you can think of that hasn't been asked about? Tell ME!! (If you think of something else later, you may certainly come back and submit it.) - no.

What did you do when the event was over? - I ran as fast as I could to tell an adult. At the time I actually tryed to convince my mother to call the police because there was a gorilla lose in the woods. I remember I was shocked by how long it took me to run back to the house. I hadnt realized that I had been so far out to the point that I had walked completely around the curve of the field and i couldnt even see the line of trees that divided the field from our back yard

Anything else you just need to vent and get of your chest??? - no

----

Enter supporting content here

contact : info@IndianaBigfoot.com

or write:

IndianaBigfoot.com

P.O.Box 420

North Webster, IN 46555

++

DATE: Monday, May 13, 2013

TIME: between 8:45 - 9:00PM

LOCATION: near Spencer, IN in Owen County

TYPE: Visual sighting

AREA / TERRAIN: rural, farm land, cow pasture, forest and woods

WEATHER: had been cool sunny day, turning into cool clear evening - no fog (civil twilight established to be approximately 9:21pm for this date - MRB)

DESCRIPTION OF EVENTS: [Where indicated by " " marks around the text, these are the actual direct quotes taken from the witness's report made to me by handwritten submission.]

"To whom it may concern:

 I think I have a "Bigfoot" sighting for you. My name is Xxxx Xxxxx, address xxx, Spencer, Indiana 47460. Phone Xxxxxxxx. Sighting at 8:45 - 9:00pm on 5/13/2013. Cool, sunny day - clear, cool evening - No Fog. Distance 80 - 100 yards away from me. No noise, no movement, horses, ducks - pigs - dogs had no reaction at all, also no smell at that time."

[she then hand drew a map of the family farm to help me visualize what she had seen - I compared this to an online Google Earth view and found it to be as presented]

"No animals close to place I believe I saw a *Bigfoot*. Sun going down - but plenty of light still. New moon visible. It was quiet though! Unusually quiet - no birds etc.

I had walked from the house to the chicken coup to close them up for the night. Turned to return to the house and looked out at the pond to see if I could spot the ducks. At the end of the pond is a hill (dam) 8 - 10' high. Scrubby trees, bushes grow on to top. The soil is hard, clay-muck mixture that was dug out of the pond and stacked up there several years ago. I had taken a few steps, looked at the pond and stopped, sucked in my breath and stared. It appeared to be a creature, medium gray in color, very heavily muscled (big pecs) wide shoulders, smaller looking head. 7 1/2 - 8' tall? There is a tree close to measure, but I haven't yet. It looked sort of shaggy. I didn't see a face - no clearly defined eyes - just suggestions of eyes, nose, mouth.

   I took 5 or 6 steps to the left/right to get a better perspective - maybe a couple steps closer. It didn't move. I stood there about a minute waiting for movement. No movement. I walked slowly around around the corner of the house - went to my car and got my <brand name> binoculars out and hurried to the back deck. I looked all over with them and didn't see anything. I did smell an unusual odor. Not horrible, just strange - I don't think I remember smelling it before. Two dogs in back yard on the deck. They didn't seem excited.

{next hand written page}

There was another person in the house, but he has the flu or severe cold. I didn't tell him. I went to the area and climbed up the pond this eve. and did not see any clearly defined footprints or find hair on any trees. There were indentations in the ground but who knows? <Name> says he rides the horse up there sometimes - not lately.

I sat on the deck this evening and waited until the light appeared the same. I retraced my steps to the chicken coup and looked again to make sure it wasn't just the trees, bushes that may have formed the suggestion of a figure standing there. And NO - they didn't. I saw something on 5/13 that was not there on 5/14. The weather conditions were roughly the same - just warmer.

<Name's> room is at that end of the house - he said the dog barked alot last night. The greenery under <Name's> window is mashed down today. Not under my window though. I didn't feel fear when I saw the possible *bigfoot*, just very startled and curious. It probably would have scared me if it were closer and didn't have a body of water between us.

My neighbor next door - <distance redacted by Mike for security> away through the trees stated a *bigfoot* looked into her bedroom window a couple of years ago. Her window is 10' from the ground as the land slants away into a ravine at the back of her house. She ran screaming for her roommate, but when they got back to her room - it was gone.

{flip page over}

I'm a <age> year old woman. I'm a <professional description> as well as a <other professional description>, <description again>. I do pay attention to details and my surroundings. I fish, hunt and am modestly aware of nature and her creatures. I'm sure there are more things in heaven/earth than we will ever come to know.

I can't say without some doubt that I saw *bigfoot* on 5/13 - but I'd give it a 97 1/2% that one came to visit our little neck of the woods.

I'd be more that happy to discuss this with someone and show them the area if requested. We can also talk to the neighbor who is convinced she had a *bigfoot* visit her house.

Thank You!

<Name and information repeated here>"

I spoke by phone in July with the witness who relayed and retold her sighting to me. Her verbal recounting of events was still as vivid and detailed as what she had written to me about in May. (I had delayed getting in contact with the witness for various reasons, for which I apologized. Every report is important - especially to an eyewitness.)

I had been putting off posting this sighting because I wasn't sure how much of the report I wanted to include on the website, seeing that the history of Bigfoot research seems to have it's ups and downs with those attempting to invade the privacy of both witnesses and those who investigate their claims. I was going to really *water down* the content of the report, but instead, I have intentionally redacted what I feel is the most specific information about the witness that should protect her privacy the most. I even considered not referring to the witness as a 'her' but as 'they'.

This general area has been a good source of activity for some time now, as much of southern Indiana, and the area west of Indianapolis running over to the western state line along the Interstate 70 corridor.

I've asked the witness that if any additional activity were to occur to let me know, or if she were to hear of any as well. There are several active researchers in this general area of the state, so if something significant were to come about, there could be "boots on the ground" very easily.

As is with any report, I can only take the witness at their word, and having spoken with her, I'm comfortable believing that she likely did see a sasquatch/bigfoot on that Monday evening on her property.

contact : info@IndianaBigfoot.com

or write:

IndianaBigfoot.com

P.O.Box 420

North Webster, IN 46555

++++++++++++++++++++++++++
++++++++++++++++++++++++++

  up on cleaning the camp ground bathrooms, in morgan-monroe state forest, a labor had report seeing a Rattle snake and look towards it, and then looked at the tree line in the boy scout ridge camp ground, and saw a 7 foot browish colored Bigfoot try to hide behind a large tree, he kept quiet about it till 2012, when researchewrs had a camp out there in June 22nd, and he told about it, also a 8 inch foot print was taken from three foot from a tent one morning in the same location, and pictures was taken of foot print.

++++++++++++++++++++++++++
++++++++++++++++++++++++++

a man named Tim, has had two sightings at the front gate to the Stepp cemetery in Morgan-Monroe state forest, he went there for a ghost hunt, and reported this account on two different occasion, some of reports were not to detailed at all, but had to be reported,

++++++++++++++++++++++++++
++++++++++++++++++++++++++

Tim , with a female assistant, was leading a group around lake Monroe, and was getting dark, and came around a bend in a trail when they heard some thing, and tim put on the flash light and pointed it towards the noise, and there standing in front of him, was a large dark hair covered bigfoot, holding its hands up to shield his eyes from the light, the female assistant back pedaled and ran back up the trail, this encounter opened the door to Tim, who is a wild life biologist, the assistant quit a week later

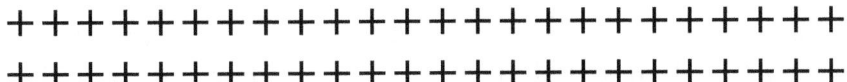

on the show finding bigfoot, comes several reports from Northern Indiana as well as southern Indiana, I suggest you all watch the Indiana episode to find these reports, there are going to be a few breaks in between the reports as the author comments, several reports will be hard to understand, the reader must understand, that the person reporting these reports and incidents, do not want to be known, because they do not want to be hassled after a climatic event, shock and emotional feelings interfere with simple thinking, and the reported must be checked by professional researchers to authenticated.

Therefore, the people's real names will remain hidden at times, but the stories are very real and absolutely true, the hidden mysteries in the forests and streams of Indiana are a thing to ponder.

++++++++++++++++++++++++
+++++++++++++++++++++++

Steve Abney was contacted by a hunter, here is his report, I am a young hunter, who deer hunts on private property every year, the property is a thickly wooded area, with Ravines and large hills, some areas, man cannot even get in to , I am a fit and healthy person and i walked a mile and a half in to the area one morning about six am, and had the feeling of being watched from the thick trees around me, I could not see anything, but the feeling persisted, and about two hours of this, I got the feeling of being in danger, I turned around and started to leave, when I heard a large growl, and then loud crashing through the timber, some big and heavy was running on two feet, this scared me so bad, I ran all the way out of the area, and will not go back.

++++++++++++++++++++++++
++++++++++++++++++++++++

walk through for sasquatch creatures, Miami Indian Lands,

Steve Abney got permission to enter Miami Indian lands,

Steve and violet Abney went to the area, and did a walk through, took several pictures of stick formations, X formations, and found old foot prints leading through the over grown woods and brush,

++++++++++++++++++++++++
++++++++++++++++++++++++

Report taken by Steve Abney,

A child of nine years old woke up one morning and rushed in to the kitchen and spoke to her mom, and blurted out the words, I have a Sasquatch family, the mother said, What do you mean, and the child told her, we have a foot print on the front cement porch, I saw them in my dream, they went out on the front porch and sure enough there was small foot print on the porch. After that they went to where the childs window was located, and by the bedrrom window the ground was tore up, where the small forest person had been pulling grubs to eat. Several days later there was two sightings at the home by the mother and daughter.

Area kept secret, pictures of the foot print was taken and hair samples collected off the back porch,

++++++++++++++++++++++++++
++++++++++++++++++++++++++

2010,April,

location,Big Walnut Creek off US 40.Parke and Putnam County,

witness, Bill???,

last name with held for private reasons,

weather conditions- warm and cloudy,

what he saw, I was driving west on us 40, and I blew a tire and pulled off the side of the road, by the bridge to change it, and was worried to getting to work late, I started to change the tire at 630 in the morning and it was still a little dark, I heard a commotion under the bridge, like a scream or howl, I walked over to the 15 feet to the bridge and looked down and saw some thing huge, with hair covered all over, running through the water, heading south, it disappeared in the trees and was gone, it only took a few seconds for this being to get out of sight, it was black in color and about 7 to 8 foot tall, massive chest area, and was running on two legs all the way, i was so scared, i sit in the car for a while and was late for work, I finally got the courage to finish changing the tire, and went to work, I did not tell anyone of encounter, til I found out that my neighbor was a bigfoot researcher, I told him my story for his reports, I still do not know what it was, but it scared me to death,

Bill told me his story, and I believed he had seen some thing, he was still to scared to talk about it, but finally calmed down enough to tell me what I needed to know, I find him sincere and a honest man,

report taken by Steve Abney

++++++++++++++++++++++++++
++++++++++++++++++++++++++

a group member and researcher, was looking through binoculars and saw a white bigfoot in a tree, the bigfoot was looking at him, he got scared and the white bigfoot left the tree, and the researcher left the area,

++++++++++++++++++++++++++
++++++++++++++++++++++++++

I was standing in line at a food bank recently, and a man recognized me and we was talking about recent reports, he was from green castle In,

 and another man behind me was listening in, and he tapped me on the shoulder, and told me his name was Doug, he asked me if he could report a sighting he had back in 1991, I said yes , of course, here is his story,

I am from putnam county Ind, I went to washington State to MT, saint Helens for a hunting trip, , we was getting bored after a long days hunt and not finding anything, so we decided to plink at some snow caps further up on the mountain, we all had high powered rifles, there was three others besides me, we was looking and shooting upwards to the west at some snow caps, and the sun was in our eyes,

all of a sudden from behind a snow cap, a huge hairy figure rose up man like, and put his arms in the air and you could see a brownish colored hair all over him, , it scared all of us and put our weapons down, and quit shooting, this figure had to be about 9 feet tall, we walked back to our vehickles and left,

I took his report and thanked him and told him I would use his first name only in the report,

date febuary 22 2011

Report taken by Steve Abney

++++++++++++++++++++++++
++++++++++++++++++++++++

Wheatfield, Indiana

I'm from Wheatfield Indiana and I saw a "bigfoot" once. November 2000 I was with a friend of mine, Jordan; we were walking through some woods near his house at night trying to scare his brother getting him to come too, but he wasn't there. We were about 10 min. from the beginning of the woods when we heard a screeching noise, it was kind of low pitched and scratchy, but it sounded close. Then while we were walking back we could see the streetlights far-off through the trees, when something went by, then we heard a noise behind us a few seconds later we both saw something crouching behind some tree stumps about 30 feet away his dad cut down a few months earlier for firewood. The thing was a dirty-gray color with a human-like face; it then ran off in the other direction when it saw us. We took off toward the road. The next day we both went out by the stump and found nothing more, no footprints no nothing.

"John Johnson"
webmaster@xxxxx.xxx.com

++++++++++++++++++++++++++
++++++++++++++++++++++++++

DATE: Monday, May 13, 2013

TIME: between 8:45 - 9:00PM

LOCATION: near Spencer, IN in Owen County

TYPE: Visual sighting

AREA / TERRAIN: rural, farm land, cow pasture, forest and woods

WEATHER: had been cool sunny day, turning into cool clear evening - no fog (civil twilight established to be approximately 9:21pm for this date - MRB)

DESCRIPTION OF EVENTS: [Where indicated by " " marks around the text, these are the actual direct quotes taken from the witness's report made to me by handwritten submission.]

"To whom it may concern:

I think I have a "Bigfoot" sighting for you. My name is Xxxx Xxxxx, address xxx, Spencer, Indiana 47460. Phone Xxxxxxxx. Sighting at 8:45 - 9:00pm on 5/13/2013. Cool, sunny day - clear, cool evening - No Fog. Distance 80 - 100 yards away from me. No noise, no movement, horses, ducks - pigs - dogs had no reaction at all, also no smell at that time."

[she then hand drew a map of the family farm to help me visualize what she had seen - I compared this to an online Google Earth view and found it to be as presented]

"No animals close to place I believe I saw a *Bigfoot*. Sun going down - but plenty of light still. New moon visible. It was quiet though! Unusually quiet - no birds etc.

I had walked from the house to the chicken coup to close them up for the night. Turned to return to the house and looked out at the pond to see if I could spot the ducks. At the end of the pond is a hill (dam) 8 - 10' high. Scrubby trees, bushes grow on to top. The soil is hard, clay-muck mixture that was dug out of the pond and stacked up there several years ago. I had taken a few steps, looked at the pond and stopped, sucked in my breath and stared. It appeared to be a creature, medium gray in color, very heavily muscled (big pecs) wide shoulders, smaller looking head. 7 1/2 - 8' tall? There is a tree close to measure, but I haven't yet. It looked sort of shaggy. I didn't see a face - no clearly defined eyes - just suggestions of eyes, nose, mouth.

   I took 5 or 6 steps to the left/right to get a better perspective - maybe a couple steps closer. It didn't move. I stood there about a minute waiting for movement. No movement. I walked slowly around around the corner of the house - went to my car and got my <brand name> binoculars out and hurried to the back deck. I looked all over with them and didn't see anything. I did smell an unusual odor. Not horrible, just strange - I don't think I remember smelling it before. Two dogs in back yard on the deck. They didn't seem excited.

{next hand written page}

There was another person in the house, but he has the flu or severe cold. I didn't tell him. I went to the area and climbed up the pond this eve. and did not see any clearly defined footprints or find hair on any trees. There were indentations in the ground but who knows? <Name> says he rides the horse up there sometimes - not lately.

I sat on the deck this evening and waited until the light appeared the same. I retraced my steps to the chicken coup and looked again to make sure it wasn't just the trees, bushes that may have formed the suggestion of a figure standing there. And NO - they didn't. I saw something on 5/13 that was not there on 5/14. The weather conditions were roughly the same - just warmer.

<Name's> room is at that end of the house - he said the dog barked alot last night. The greenery under <Name's> window is mashed down today. Not under my window though. I didn't feel fear when I saw the possible *bigfoot*, just very startled and curious. It probably would have scared me if it were closer and didn't have a body of water between us.

My neighbor next door - <distance redacted by Mike for security> away through the trees stated a *bigfoot* looked into her bedroom window a couple of years ago. Her window is 10' from the ground as the land slants away into a ravine at the back of her house. She ran screaming for her roommate, but when they got back to her room - it was gone.

{flip page over}

I'm a <age> year old woman. I'm a <professional description> as well as a <other professional description>, <description again>. I do pay attention to details and my surroundings. I fish, hunt and am modestly aware of nature and her creatures. I'm sure there are more things in heaven/earth than we will ever come to know.

I can't say without some doubt that I saw *bigfoot* on 5/13 - but I'd give it a 97 1/2% that one came to visit our little neck of the woods.

I'd be more that happy to discuss this with someone and show them the area if requested. We can also talk to the neighbor who is convinced she had a *bigfoot* visit her house.

Thank You!

<Name and information repeated here>"

 I spoke by phone in July with the witness who relayed and retold her sighting to me. Her verbal recounting of events was still as vivid and detailed as what she had written to me about in May. (I had delayed getting in contact with the witness for various reasons, for which I apologized. Every report is important - especially to an eyewitness.)

I had been putting off posting this sighting because I wasn't sure how much of the report I wanted to include on the website, seeing that the history of Bigfoot research seems to have it's ups and downs with those attempting to invade the privacy of both witnesses and those who investigate their claims. I was going to really *water down* the content of the report, but instead, I have intentionally redacted what I feel is the most specific information about the witness that should protect her privacy the most. I even considered not referring to the witness as a 'her' but as 'they'.

This general area has been a good source of activity for some time now, as much of southern Indiana, and the area west of Indianapolis running over to the western state line along the Interstate 70 corridor.

I've asked the witness that if any additional activity were to occur to let me know, or if she were to hear of any as well. There are several active researchers in this general area of the state, so if something significant were to come about, there could be "boots on the ground" very easily.

As is with any report, I can only take the witness at their word, and having spoken with her, I'm comfortable believing that she likely did see a sasquatch/bigfoot on that Monday evening on her property.

contact : info@IndianaBigfoot.com

or write:

IndianaBigfoot.com

P.O.Box 420

North Webster, IN 46555

++++++++++++++++++++++++
++++++++++++++++++++++++

a clear picture os a sasquatch was taken in paoli Ind and the crew from tom biscard iinvestigated it,

++++++++++++++++++++++++
++++++++++++++++++++++++

Warrick County, Indiana

1985

I live in Warrick County Indiana. I was across from my house fishing in a lake that we always fished in. It was in a wooded area that was across the street from a county park. The whole

area is scattered with wooded lots and farmlands.I was fishing there for about 20 minutes when I got an uneasy feeling that I was being watched or something.

As I started looking around from my right to left, I noticed something large and hairy standing in the water looking directly at me. It didn't make any movement or sound, just stood there staring at me,didn't make any noises or moves at all..

I dropped everything I had and ran up the hill, through the woods and across the street.

I was about 70 yards form my house and it all accrued at about 4 pm in the summer time. I don't care that my friends and family think I am crazy, I will go to my death bed knowing that I was looking at bigfoot. He was "roughly 6-7 feet tall" - he was in water up to his knees. Dark colored fur and that was about all I got a look at before I bolted for home.

I will never forget what I saw. It scared the hell out of me, I refused to go into those woods for many years, and I wouldn't go alone, not after that. If you need any more information,

Contact me at yackumhead@xxxx.xxx, not the email I sent this from. Thanks Jason Tremper

Report filed with Bobbie Short, Saturday, November 03, 2001 11:36 AM

(E-mail address in the database)

++++++++++++++++++++++++++
++++++++++++++++++++++++++

Newport Indiana 1985

In 1985, I was attending Indiana State University in Terre Haute, IN. One Saturday morning I was travelling to Chicago to visit relatives. I was driving north on U.S. 41 and I looked to my right down into a small valley (more a depression in the land than a valley) with a small stream running through it. About 100-150 yards from the highway, I saw a large dark-colored creature walking away from the road (to the east.) It was very broad and I could clearly see the creature's arms swinging. My view of the bottom part of it was obscured by tall grass and I could only see the very tops of its legs. The head never turned at all. I saw this for 4-5 seconds before the trees on the north side of the depression obscured my view. There were also trees on the south side. My immediate thought was that I didn't really see what I thought I did. I continued driving, unsure of what to do. After a few minutes, I turned around and drove back past where I had seen it (my view from this side of the highway was blocked), turned around again, drove to the bridge where the sighting occurred and stopped where I had originally seen it. There was nothing there at this time. Not being aware of any sasquatch sightings in this area (or the Midwest in general), and not knowing what else to do, I continued on to Chicago and didn't think much more about it for many years.

http://www.untoldthemovie.com/newportin.html

++++++++++++++++++++++++++
++++++++++++++++++++++++++

Morgan County, Martinsville , Indiana

Morgan Monroe State Forest 1984

A college student attending Indiana University in Bloomington, Indiana was living a small house thats property line bordered morgan- monroe state forest. Since he had moved in he spent little time there during the day and basically it was a place to sleep. But sleep was getting difficult because of the sounds that would awaken him that came from the forest. The high peircing screams would usually send him out of his bed and running for the nearest light.

Friends that would visit on the weekends said that they saw something running across his driveway behind their cars in their rearview mirrors. Rocks were being thrown on his roof at night and the constant pelting sound made it impossiable to sleep.

Many times he could hear the sound of something walking around his home accompanied by the sound of loud breathing. He would walk outside look into the darkness but he saw nothing. So he decided to order a pair of night vision glasses from army surplus.

Everytime he would try to use them the sounds he was hearing would stop. It was getting obvious whatever he was hearing could see the beam from the night vision glasses.

Then one night he got home very late and he had the night vision glasses in his backseat, for some reason he picked them up turned them on and started looking through them.

The first thing he saw was standing in his driveway about twenty feet away and was staring directly at him. It was at least 7 feet tall , large head , no neck and covered with long dark hair or fur. Its arms hung to its knees at its sides and it stood motionless. What he described as a 'diamond shape pupil' that he could see thru the glasses, shined in the light of the one outside security light. He just stared at it and he did not move, then it turned around took two steps and stepped behind a tree, and it did not move. What he noticed was the color of the hair was darker between its shoulder blades and down its back. The young man could still see the side of it. It seemed to be waiting to see if it was being followed. The man backed up and walked into his house not taking his eyes off of the hairy shoulder that could still be seen from behind the tree.

++++++++++++++++++++++++
++++++++++++++++++++++++

parke county Ind, 1975

four teenagers musroom hunting, went back to car, saw 8 foot black bigfoot coming towards them on gravel road, got scared, got in car, and tried to start car, it was flooded, when bigfoot got up on the car, the car started, and back fired, making the bigfoot mad, and the creature hit the trunk of the car with both hands, kids took off, and went to sherrifs office in brazil Ind, and cops claimed the kide destroyed the trunk with a sledge hammer, making fun of them,

++++++++++++++++++++++++++
++++++++++++++++++++++++++

La Porte County, Indiana

Kingsbury Ordinance Plant, October 1987

In October of 1987 a friend and I went camping just outside of Kingsbury Ordinance Plant (KOP for short) in Kingsbury, IN. The nearest city there would be Stillwell. We were about a mile to a mile and a half off Hupp Road (which is the road the enters KOP).

We set up camp just before dark around 6 pm or so. After about 45 min, my friend went home to eat and call another friend of ours to come out when he gets off duty. Just around dusk, while getting fire started, I heard rustling in woods about 25 yards away. I didn't think anything of it at the time thinking it might be a deer. After about 10 minutes, heard noise again, but it was about 15 yards away this time. I yelled "Hey" real loud and the noise stopped for about 5 minutes.

Then I heard grunting about same distance away. Getting nervous, I yelled again and said, "Knock it off." I then picked up a stick and said that I had a gun. The noise stopped after 3 or 4 sec and as I turned to look at fire, I heard a loud thump to my right.

I looked and saw a large tree stump that landed about 8 ft away. I turned around and faced the woods and yelled very loudly, "Knock it off or I will shoot!" I then saw another large log sailed through the air toward my direction and land about 3 feet in front of me and about 5 feet to left.

I started running down a trail toward the road. I did not hear anything for about the first 100 yards, and then I heard rocks moving as if someone was running in the same direction as I, but they were on the RR tracks to my left.

The sounds of the steps were about 1 sec. apart. I heard this for about 8 to 10 sec. then it stopped. I ran for about 25 more yards and stopped. It was dark and overcast skies. I turned and looked behind me and then to my left. I then turned back toward the road and walking very fast, I came to a dip in the trail. This is where I saw a large figure standing in front of me. The dip in the trail sloped down to about 3 feet deep and ran the depth for about 30 feet then it rose again to ground level.

The figure was standing in the center of the dip in the trail and stood at the same height as myself. I am 5 ft 9 inches; so I figure it was from 8 to 9 foot tall. It was dark, but it stood about 8 ft tall and was dark and hairy. This thing let out a scream that would make your blood curdle. I did what any normal person would have done; I took off running toward my right through an open field. I ran as fast as I could away from this figure, until I finally reached my friends front steps about ¾ miles away. At first I thought it was him, but when he came to door he had been eating dinner, his shoes where off, and his mom said that he had not left the house. After he ate and our other friend arrived we all went back out there about 3 hours later. The small tent we put up was tore down and logs and stump was lying in the spots that they landed. Whatever I saw was not human. It was not my imagination. And to this day I will not go out there because of fear. I went out there one time after to drive through the ordinance plant and I felt that it was watching me from the woods.

Darren Bailey

Report logged by Bobbie Short, Saturday, March 17, 2001 3:03 AM

++++++++++++++++++++++++
++++++++++++++++++++++++

Knox County, Indiana 1977

My encounters, there are four; they're old though, but I'll relate just a few facts without all details:

1)

Mid September 1977,

Two friends and I were walking to the river to check some throw lines, when we saw an old one take off running, (I say it was old because it's coloration was a white/grey). It was running on two legs fairly fast, it was dark, but we had flash lights and must of startled it, I suspect it was hunting. It looked like instead of a neck, it had a hump on it's back. Initially, we just thought it was just someone pulling a prank on us, so we took off running after it, but lost sight of it because it out ran us, crossed a high fence and vanished in the woods.

2)

Mid October 1979, two different friends and I went out predator hunting at night. We would pull off the roads at likely looking spots, stick speakers out and play a rabbit in distress call, while waiting beside the car with flashlights and guns. At one spot we called in one that yelled/roared at us, its eyes glowed yellow, and it must've been younger, cause it was black. After it yelled again, when I shined my brighter light on it, (about 50 yards away), we shot at it and it sure busted brush getting out of there. It really shook me up.

3) My cousin and I went fishing, about mid June 1980, and heard one across the old bed from us right before dark, breaking sticks and just staying inside the foliage out of sight, it later crossed the old bed onto our side, it kept making this moaning noise getting closer and closer, until we could hear it in the horse-weeds right behind us, we were freaked. Then I heard my mother's car coming down the gravel road, we then heard it move off fast, but not as fast as we did!

4) Late Sept. 1980. I had my mom let me out at in a woodsy place before daylight so I could squirrel hunt, (not same place as June). I walked into the middle of woods and sat down to await light. Just as it was starting to get light enough I heard something moving my way through woods, when I heard it moan, I just got my 22 cal. ready but

froze against tree. It passed by me at less than 50 yards.

It never even knew I was there! I guess I was lucky for the crosswind. It was also a younger one, (black), As soon as it got out of hearing I went the other direction, I never even hunted that day I just waited at the edge of woods until my mom came. (note): This last place is the EXACT same place, 20+ years later that this girl had her extremely close encounter. I've already made steps to contact her, but have not done so yet. All these encounters are within 3 miles of one another.

I am older and bolder now, so I hope to gain more insight, contact, or kill it if necessary, (if it attacks me), I just want to know, I've GOT TO KNOW!
THANKS,

P.S. PLEASE, do not relate these to any one in a way that will pinpoint my location or me; I don't want alot of people tramping through my areas, perhaps scaring them off.

I wouldn't mind just one like-minded S.W. Indiana resident to help, and any first-hand information on how I should approach this would help. I've considered and bought night-vision, a semi-auto 30. cal. rifle, predator-calling tapes, and books, videos, and researched your web sight. I'm also planning on buying night-vision camera, and a cam tracker, and have just bought cast-making plaster, this will be a part-time thing cause of my work and coaching kids baseball, but I am determined.

The worse part is not really knowing more than what I do, but I guess that's also my motivator; That and the fact that people think I'm nuts if I mention anything about it. Any help will be deeply appreciated.

- ---

Exact location not to be made public or informant name revealed.

Information is held private by informant request and contained in database files.

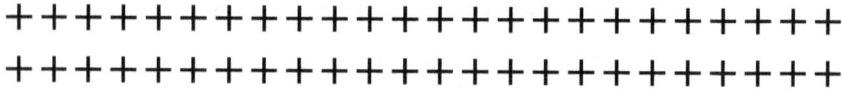

Gibson County, Indiana

Nearest City, Princeton, July 1995

Early afternoon near the lake at Camp Carson

I am writing this for my son. He has told thve same story repeatedly. My son was away at camp. He was at Camp Carson in Princeton, Indiana. We live in Evansville, Indiana. The sighting took place in 1994 or 1995.

His story: The group was trail hiking in very tall grass about 3-5 feet high. He was at the back of the group. He stopped to tie his shoes and the group was out of sight by the time he was done tying his shoes. When he stood up he realized the group was out of sight. He started to jump up to see if he could see where the group was.

This is when he saw something very tall with long black hair, with an oval shape head. It was standing up with his back to my son and he appeared to be watching something in the surrounding trees.

He was about 6-7 feet tall, long black hair, but wasn't able to see face because the creature's back was facing him. Tristan started to cry because he was scared. He ran down the path screaming for the others.

He ran until he found the group and then told the female counselor his experience but she refused to believe him. Tristan has been back to Camp Carson several times since his encounter, but refuses to go trail hiking.

From: Tristan Madison's Mother.

Sunday, November 10, 2002 at 09:48:13

REMOTE_HOST: cache-rm04.proxy.aol.com

REMOTE_ADDR: 152.163.189.132

Report logged by Bobbie Short (Map below)

++++++++++++++++++++++++
++++++++++++++++++++++++

http://www.bigfootencounters.com

Bartholomew County, Indiana-

August 15, 2006

A young man was squirrel hunting in a remote area about 3 miles from his home. He had created himself the network of trails that he frequently traveled to reach his favorite places in the forest and he knew it very well.

In the past he had seen several footprints that he had no idea where they had originated but he had not given it very much thought. He had found a place to use his game call that he also could conceal himself.

After several minutes of using his call he was startled by a very loud whooping call, very similar to the Puyallup, Washington recordings.

When he looked in the direction of the sound he could see it running thru the woods basically circling him. It stayed on all fours but he said that it was still huge. Fur or hair was black in color. He told me even though he had a 12-

gauge shotgun he was terrified and he had no inclination of trying to shoot it.

He ran out of the woods backwards, he would not completely turn around to run and he did so for 3 miles. The creature followed him for at least half the distance home. He said this is the second time encountering it the first was a couple of weeks ago but then he only saw the glowing red eyes and nothing else.

++++++++++++++++++++++++
++++++++++++++++++++++++

Dearborn County, Aurora, Indiana

Published in the Cincinnati, Ohio Post April 20, 1977

Tom and Connie Courter were coming home at night...as the husband got out of the car, a monster collided with the car..

The next night when they came home they saw it perched on the hill. Tom Courter shot at it 15 times with a .22, describing it as a hairy, apelike creature approximately 12 feet tall. It made a noise like "ugh."

++++++++++++++++++++++++
++++++++++++++++++++++++

Allen County, Indiana

Near Fort Wayne July / August 1985, evening

Don't know if it was Bigfoot or not, but about July or August of 1985, in the evening hours - about twilight, my brothers and I were walking down the railroad tracks behind our house in the direction of Fort Wayne - Pretty large city for a sighting of bigfoot, but non-the-less, I will tell you what we saw.

As we walked, we talked and threw rocks and were generally being young boys - (I was 12 at the time) when my oldest brother noticed something that we had never seen before - and being that we always walked these tracks, we knew if something didn't belong... At any rate, it looked like a large mound of black dirt just off the left side of the tracks, fairly near the rails. We couldn't really tell what it was in the evening light, but we continued to walk toward it anyway, talking and

kicking stones, not thinking a whole lot about it. We got probably within 100 to 150 feet from it, when quite to our surprise, it stood up on two legs - quite big and very tall I might add - maybe 7 to 8 ft. in height, and with two huge steps it cleared a double set of railroad tracks and ran down the ravine on the right side of the tracks and disappeared into the woods. It was tall, 7 to 8 feet, stood up on two legs; massive upper body.

We ran like hell for home and told our dad who reassured us it was probably just a deer - I accepted that at the time - even though I have yet to see a deer stand up on two legs and sprint across two sets of railroad tracks. To date, I cannot claim it WAS bigfoot - we never got a close look at it, nor did it let out any "common" bone-chilling yells or other sounds. But whatever it was, it wasn't an animal we had ever seen before – perhaps it was just a man - but if it was, he should have been in the record books.

The area has railroad tracks - slight hills, grassy and small bushes on the left side. Wooded with a slight ravine on the right side. There was a small creek and a small pond in the woods to the right side of the tracks. Our father was notified.

Sunday, October 19, 2003 8:07 PM (Map Below)

Logged in the database by B. Short

++++++++++++++++++++++++
++++++++++++++++++++++++

YEAR: 1989/90

SEASON: Summer

MONTH: July

DATE: 07-01-1990

STATE: Indiana

COUNTY: Wells County

NEAREST TOWN: Bluffton

NEAREST ROAD: State Route 216

OBSERVED: This happened at Oubache State Park near Bluffton Indiana either the summer of 1989 or 1990. I was working as a night watchman at the front gate and had been watching television when I decided that I needed to get something out of my Ford Bronco. The area leading up to the park is heavily forested with trees that had been planted by the Civilian Conservation Corps during the Depression. The road leading into the park was about a third of a mile long.

As I stepped outside the booth and was crossing the road to the parking lot, I heard the most terrifying scream that I have ever heard in my life about 250 yards to the southwest of my location. At the same time that I heard the scream I also heard a tree snap and break and fall to the ground. The tree sounded to be quite large. The weather that night was very calm and there was no wind at all which could account for this tree falling down. I quickly turned around and headed back for my booth and sat down to think about what just happened. I felt like I was in a daze for a few minutes as I was trying to think what a logical explaination of this event could be. I sat there for about 45 minutes and after I had settled down a bit, I decided that it must have been my imagination playing tricks on me and I proceeded to head back out to my truck. I was walking toward my truck and had gotten to the exact same spot that I was in previously when I heard the horrible scream when I heard it once again and this time it was about 50 yards away from my booth. I once again quickly retreated to my booth and sat there until daylight. Once day light arrived I walked out in the woods to investigate but found nothing out of the ordinary. I didn't walk far enough out to find the tree as I was still quite unsettled by what I heard. I have heard screech owls and bobcats but nothing, absoultely nothing, that compares to the sound that I heard that night. I can't say to this day what it might have been, but this has bothered me for about 16 years and I still think of it evey now and then. Reading about some of the other sighitngs that people have heard and the sounds that they said this creature made makes me think of my own experience and I wonder if I too actually had an encounter.

ALSO NOTICED: No

OTHER WITNESSES: No

OTHER STORIES: No

TIME AND CONDITIONS: Around 3:30 or 4:00 in the morning

ENVIRONMENT: Heavily wooded State Park in Northeastern IN

Follow-up investigation report by BFRO Investigator Eric Lester:

I spoke to the witness about his experience. He describes himself as an outdoorsman -- very familiar with common wildlife sounds. Both vocals he heard were humanlike, very loud, and very guttural, and lasted about 3-5 seconds each. When he heard the first howl, a tree snap and crack was heard coming from the same area, roughly 250 yards away. He was about 25 feet out of his booth when the howl occurred.

When he exited his booth a second time, about 45 minutes later, another scream/howl was heard, which occurred when he approached the same spot that he was in when he heard the first howl. This one sounded much closer, about 50 yards away. Nothing else was experienced that night, or found the next day upon searching the immediate area.

The state park is around 1000 acres, with the Wabash river bordering its southwest side. It is a very rural with areas of dense forest.

++++++++++++++++++++++++
++++++++++++++++++++++++

YEAR: 1989

SEASON: Summer

MONTH: August

DATE: August 13th

STATE: Indiana

COUNTY: Wayne County

LOCATION DETAILS: US 4o to Salisbury road south, approximately 5 miles down Salisbury road in the woods located on the west side of Salisbury road south. We still do a lot of camping in this area and have never heard the noise since.

NEAREST TOWN: Centerville, Indiana

NEAREST ROAD: US 40

OBSERVED: August 1989 My buddy and I do extensive camping in this area and were raised here. We have heard many strange noises but on this particulary night we were camping in an old corn crib and around 2 pm we heard a very loud growling, hollering noise that woke both of us up from sleep. We both looked at each other in disbeleif but decided not to let that bother us and go back to sleep. Within a few minutes we heard the same noise again, only this time louder and much closer.

We have heard coyotes and screach owls but we have never anything even remotely similar to this sound. The next day while sitting safely in our homes we discussed what we had heard and thought about big-foot but who had ever heard of a big foot in Indiana?

Jay Smith

Mike Davis (2001)

ALSO NOTICED: Everything seemed normal prior to our night time visit.

OTHER WITNESSES: There were just the two of us and we were both sound asleep.

OTHER STORIES: Around 1976 a local family were all riding their dirt bikes in the woods and one of the bikes became disabled so everyone stopped to work on the bike at which time they saw a creature watching them from a distance that appeared to be tall and hairy.

TIME AND CONDITIONS: occured around 2pm at night

Extremely dark (noise was in the woods we could not see anything only the noises we heard and the hair on the back of her necks standing.

ENVIRONMENT: The environment consists mainly of hardwoods with intermittent corn and soybean fields. The woods are mostly undeveloped due to deep ravines. Small

areas of the woods are a little swampy, intermittent patches of pine woods.

Follow-up investigation report by BFRO Investigator Jim Osborne:

On talking with the witness he stated that the sound was so loud and disturbing that they decided to leave immediately. They only had one gun, so one held the gun while the other quickly dressed, and then they left.

Since posting the report witness has not heard of any recent sightings in the area.

++++++++++++++++++++++++
++++++++++++++++++++++++

YEAR: 1981

SEASON: Winter

MONTH: December

STATE: Indiana

COUNTY: Washington County

LOCATION DETAILS: around 1.5 miles south of South Boston or S. Rd. 160. From Mulls store go up the big hill, turn left at the first turn. About 1 mile turn left again and park at the bottom of the hill near the creak before the turn. facing north the sound came from a briary grown up area that has since been buldozered out.

NEAREST TOWN: South Boston (Salem area)

NEAREST ROAD: Olive Branch Rd.

OBSERVED: In 1981 my brother-in-law and I were racoon hunting with three good dogs in the South Boston, Indiana 47167 area. We were avid coon hunters at the time because I was laid off from my job. Well one night during hunting season we let the dogs out at one of our favorite spots and very soon the dogs hit trail. They ran across the road the wrong direction to what I wanted and they ran for about a minute. Then we heard the loudest most horrible scream I've ever heard. It lasted maybe fifteen seconds. The volume was greater then any human or any bobcat, bear, coyote, fox or anything that lives in my area. My three good dogs went silent and came running back and my best dog stood between my legs looking in the direction of the noise. I said "Roger, what the heck was that?" Roger said he had no idea. I told him to load the gun, which we never carried loaded because of fear of an accident, and he fiddled around until I took it and loaded it myself. Meanwhile my dogs were bumping into my legs and still looking into the woods. Well we slowly walked backwards to the car and instead of loading the dogs into the trunk they jumped into the back seat before I could stop them weather I wanted to or not. That ended my coon hunting for the night. Now the howl you have on this site named publicklam.wav might sound like a baby one compared to what I heard. The sound I heard had much volume and much force.

ALSO NOTICED: 1986 - I heard it again while sitting at home reading. Around midnight, I heard the dogs on the porch. They were trying to get in the door. I opened the door to see what was going on and the same noise was across the highway. Dogs were trying to get in the house.

OTHER WITNESSES: Only my Brother-in-law

TIME AND CONDITIONS: It was around 9:00 PM and was cool maybe cold.

ENVIRONMENT: Many acres of woods and pine.

++++++++++++++++++++++++++
++++++++++++++++++++++++++

YEAR: 1997

SEASON: Summer

MONTH: August

STATE: Indiana

COUNTY: Washington County

LOCATION DETAILS: near North Cave River Valley Rd and West Cave River Valley Road

NEAREST TOWN: Campbellsburg

NEAREST ROAD: near North Cave River Valley Rd and West Cave River Valley Road

OBSERVED: I have had two possible "Bigfoot related" incidents. I've wanted to discuss the older incident for quite some time but have hesitated to do so. The second incident occurred March 16, 2009 and unnerved me enough to go ahead with these reports.

(Note: see report #25658 for the more recent incident)

The first incident occurred in approximately August of either 1997 or 1998, I can't recall offhand the exact year any longer. This happened on private land near Campbellsburg, IN. The terrain of this private land is moderate hardwood forest, heavy leaf litter and only mildly rugged. There are several caves (all of them quite extensive as privately held cave systems go). There is also a valley with a small creek and a disused cabin. We had two separate incidents. On the first night we camped at the top of the valley so we could look down on the cabin and creek, mainly because it was a very pretty overlook and we could see the stars. After dinner and a couple of beers (no one was intoxicated by any means), one by one we headed off to use the restroom downwind and far away from camp. On one side of camp was the valley and on the other side was a very steep incline leading down to a privately owned field that no one could trespass without risking at least arrest, if not worse. As I started out into the forest to relieve myself, I heard the distinct sounds of someone walking purposefully up this steep incline. The footfalls were heavy enough to be heard distinctly apart from the shuffling of leaf litter. I squatted down, thinking it was the adjacent property owner coming over to complain about noise or trying to chase us off (even though we had full permission to use this property). The footfalls stopped at the top of the incline and then, whoever this was, took off running full bore into the deeper parts of the forest. The footfalls were very distinctly bipedal and NOT the rapid sound of deer footfalls. I didn't know who this may be so I called out to them. The running continued and no response was given. After I did my business, I went back to camp and the other persons agreed they heard the footfalls and thought it was me. None of us saw anything at that time. About 2 hours later we heard a VERY loud scream, starting at a low pitch and rapidly increasing to a piercing screech. We heard this only once and it caused the forest sounds to stop -- about 2-3 minutes later the normal forest sounds started again. We then decided to turn in for the night -- we had been creeped out enough. We slept in a van on our trips so we climbed into the van and fell asleep. About 2 or 3 AM, one of the other persons on the trip yelled out and woke us all up. He'd awakened to see a large arm (he never specified anything other than it was very large and the hand was large and dark) reaching in the van window and taking food out of the front seat. We looked and the food was partially missing, the remainder spilled about the seat and floorboard. I can't say for sure if he was dreaming, hallucinating or just spooked from earlier and mistook a raccoon for an arm or something like that. Something had disturbed the food but I don't know what it specifically was -- though I personally made the connection to our "visitor" from earlier visiting the camp. No tracks were found in camp but we could see where something large had ascended the steep incline the night before when we investigated the next morning. These were just gouges in the hillside and disturbances in the leaf litter, nothing more. But whatever or whoever came up that incline had marched right up and never hesitated. We didn't have tools to measure, photograph or record what we saw as we didn't think it was "bigfoot" but rather an animal or the other property owner at that point. We camped in the valley for the remainder of the trip without incident.

We returned about a year later. We decided that we'd camp in the valley for the trip -- it was creepy on top of the hill plus we were still thinking the adjacent property owner would be angry again at us being up there with a bonfire and making noise at night. We had 3 uneventful nights in the valley. The fourth night was different -- other than the noise of the stream, the normal "night sounds" were generally absent, except for the occasional owl calls or other sporadic noises. It was quite eerie that night. We went to bed around 1AM (this time we were camping in tents and not in the van). About one hour later, we heard the familiar footfalls, this time descending the valley wall on the opposite side of the creek. There would 2 or 3 footfalls, a sliding sound then 20-30 seconds of silence. This process repeated until (we assumed) the person reached the edge of the creek opposite us. After what seemed like an eternity, the person crossed the creek very slowly -- we could hear gentle splashes with each step. Once in our camp, we could hear shuffling sounds. Our fire had gone out, so we could not see anything through the tent walls. None of us really cared to confront whoever or whatever was visiting. After about 30 minutes of shuffling, our cooler was forcefully knocked to the ground. Silence followed and we all fell back asleep. I awoke sometime later to the sound of someone approaching the tent. As I looked up through the rainfly, a head appeared and looked down at me. As we were under the tree canopy, the lighting was very poor. I could see that the head was very large, there was coarse hair on top and sticking out to the sides in places, but the face was in shadow. At that point I gasped and the head quickly moved out of view. I could hear it cross the creek and head up the hill and it was gone. I mentioned this to the guys the next morning and none of them noticed it return for the second visit but me. We had no further disturbance.

Neither time did I feel "threatened" or any negative feelings other than the fear of encountering an unknown visitor, whether human or otherwise. But the head looking in the tent, and the stealthiness of the visitor, clued me in that this was not a "man" looking to scare us but rather likely a startled Bigfoot (the first encounter) and then an inquisitive Bigfoot (second encounter). I poked around quite a bit on the property and didn't see any obvious sign of a large animal, but there were lots of deer and the disused cabin had several bedding areas in it. Something or someone had bedded down there many times in the past, but the areas looked disused as well. Also note these bedding areas were on both the first and second floors of the building.

ALSO NOTICED: Nothing out of the ordinary either time

OTHER WITNESSES: 4 witness, no contact with any of them now

ENVIRONMENT: First incident on private property -- moderate hardwood forest and also a valley/creek system

++++++++++++++++++++++++
++++++++++++++++++++++++

YEAR: 1999

SEASON: Summer

MONTH: June

DATE: 12

STATE: Indiana

COUNTY: Washington County

LOCATION DETAILS: From SR 135, go to Dutch Creek Rd and from there, turn on to Falling Creek Rd.

NEAREST TOWN: New Pekin

NEAREST ROAD: State Road 135

OBSERVED: My sighting of what I believe to have been a sasquatch was quite brief, though I did get a decent look. My mother and I were visiting my aunt and I was asked to go out to the car to get my mother's cigarettes. I went outside and walked over to the car and grabbed the cigarettes. When I got back out of the car and went to close the door, I noticed something moving in the field next to my aunt's property. A very large and definitely bipedal animal was walking down the field towards the woods. It walked with a long and sort of slouched gait. It was either covered with really dark brown or black fur and had rather long arms that swung loosely as it moved. It crossed the distance of at least 200 feet in less than 20 seconds. As soon as it reached the woods, I walked into the house and must have sat for at least half an hour trying to process what I'd seen. I was only 13 at the time.

ALSO NOTICED: In the fields close to the area, I'd noticed several spots where something had bedded down. I just figured it was deer, but I did notice an unpleasant odor around them once or twice. Also, since reading through this and other sites concerning bigfoot research, I did notice some things in the woods. Groups of large sticks stacked in like a tent fashion. I just thought it was the kids in the area doing it, it could still be that.

OTHER WITNESSES: No other witnesses, I was only going out to the car to fetch something.

OTHER STORIES: I did hear that a couple years later some kids said that they were followed by a bigfoot near the woods just 1/4 of a mile down the road from where I had my sighting.

TIME AND CONDITIONS: The time was in the early afternoon and it was a clear, sunny day.

ENVIRONMENT: Overgrown field directly off to the right of the road. There was a small grove of trees near the field. Also across the road from these trees there was a few acres of woods with at least one small creek running through them.

Follow-up investigation report by BFRO Investigator Stan Courtney:

I spoke with the witness by phone. The distance was over a hundred yards but the witness stated that he could clearly see the animal.

The animal was dark brown in color, seven feet tall and very heavily built. It appeared to have weighed about four hundred pounds. The animal did not turn and look at the witness but just walked through the field and back into the woods. He was never able to see the face.

++++++++++++++++++++++++++
++++++++++++++++++++++++++

YEAR: 2006

SEASON: Summer

MONTH: May

DATE: 30th

STATE: Indiana

COUNTY: Washington County

LOCATION DETAILS: Fredricksburg,IN

NEAREST TOWN: nearest larger city New Albany, IN

NEAREST ROAD: highway 150

OBSERVED: I have been reluctant to beleive what my mother says because she has always been a little squirrley, it runs in her family, but she told me several years back that my uncle was outside one night drunk as usual, another reason i was reluctant to believe her, but the story goes that something big and hairy got after him and he crawled up under the pickup truck to keep it from getting him. They live in southern Indiana in a remote area. They own about 60 acres that was my grandmothers. The area is rolling hills some big enough to be considered mountains. It lies northwest of Louisville, KY and east of Paoli Peaks, IN, a fairly well known ski resort town. I often wondered about the things they say they have seen and heard in the woods. They live about 3/4 of a mile back off the road in a trailer in the middle of a thick woods with alot of tall cedar and pine trees oaks and walnut trees also. the woods is heavy with alot of undergrowth and alot of sink holes and caves. Houses are spaced out in the area with alot of other heavily wooded properties joining theirs. Years ago my grandmother didn't know what to call it but she reported to us of being chased back the lane by 2 adults - a larger adult and a smaller adult and 2 young ones. My siblings have heard howls but not really seen anything. My brother was killed in an automobile accident on Memorial day and we went up to help them make funeral arrangements on May 30th 06. I was on my way to the truck, it was probably between 11pm and midnight, my kids and husband were already in the truck we all had colds and my son was caughing so he my daughter and my husband did not hear it. It was a howl that made the hair on the back of my neck stand up. I am a country girl, I've heard a coyote, dog and wolf howel before and this wasn't any of those. I listened to the howls on your website and that is exactly what I heard only this one was alittle higher pitched like it may have been a younger one. My husband said all he could hear was my son coughing, I didn't want to be the only one who heard it. I told my sister the next day and she said yes I've heard it many times before. You don't have to prove to me that they exist I just want to know if that is what is in their woods. I am concerned for my parents safety. They are both dissabled and don't need to be back there alone but they are too stubborn to sell it and get closer to town and when they are gone that land will go to us and I don't intend on sharing it with "Old Hairy." He's gonna have to pack his family and move on. I'd like to have proof to show the world that yes, they do exist. [Edited]

OTHER WITNESSES: several over the years, grandmother who is now in nursing home with alzheimers my mother, my uncle my dad doesn't talk much so i'm not sure what experience he's had. I only heard the one howl, but my sister has heard it many times.

OTHER STORIES: I would like to know if any other people from this area have heard or seen anything, but the thing is it is a low income rural area with alot of druggies and drunks and your basic scum of the earth type people that you either may not trust or would have trouble believing. There are a few reputable people in the area but i don't live there and have the time to try to find much of anything out.

TIME AND CONDITIONS: 11pm to midnight pitchblack slight fingrnail moon hot muggy

ENVIRONMENT: alot of tall trees oas hickory and others heavy undergrowth alot of cedar and pine very hilly area land not real good for farming so it's mostly either grown up wooded or pasture.

Follow-up investigation report by BFRO Investigator Eric Lester:

I spoke to the witness about what she experienced. When she heard the howl she was outside of her truck and heard what at first sounded like an ambulance. She eventually realized she was not hearing an ambulance, but a howl. She compared it to the Ohio howl found on the website, but said it was higher pitched and lasted a little longer. It also sounded as if was far away. Her family was in their truck at the time and could not hear it. She said it isn't similar it to anything she's ever heard before.

The story she heard from her grandmother was when she was young, so no details were obtained at the time. She also stated that other members of her family have heard howls like the one she heard - her sister has heard them on multiple occasions, and her daughter has heard something she said 'was like the dinosaurs in the movies'.

The area where this occurred is close to a mile down a rural dirt road, surrounded by thick woods on all sides. There is some farmland in the area, and lots of natural caves. The Blue River runs through Fredericksburg, which is located between Hoosier National Forest and Clark State Forest.

++++++++++++++++++++++++
++++++++++++++++++++++++

YEAR: 2006

SEASON: Summer

MONTH: July

DATE: 2

STATE: Indiana

COUNTY: Warrick County

LOCATION DETAILS: Take I-64 to wheatonville rd and cross on to North road, you see a bunch of trucks parked on the side of the road, the incident happened about 5 miles into the woods.

NEAREST TOWN: Elberfeld

NEAREST ROAD: North Road

OBSERVED: Some neighbors and I went four wheeling on our dirt bikes and quads. It was about 6:30 PM when we left. About 9:30 PM we stopped to repair a chain on one of the bikes. We were all gathered around his bike when we all heard a moan that sounded like a cross between a tornado warning siren and wolf howl. We all froze while we listened to this thing moan about 5 or 6 times. The lead rider in our group grabbed his rifle out of his case. He hikes out into these woods all the time and I've never seen him grab a weapon to defend himself against anything in these woods.

After the moans subsided we waited frozen there for about 30 seconds when we heard about 6 knocks like someone hitting their knuckles against a tree. We were all frozen stiff in fear. Nobody made any movements. After we heard the knocks we started hearing something run through the woods. It sounded like it was running really fast. As soon as we heard this we abandoned the bike and jumped on our vehicles while Mark rode on the back of my quad on the way back.

We went back to that spot the next day around noon and the bike was not on the trail where we left it. It was about 5 ft off the trail into the brush, as if it had been tossed there. We never found any footprints or anything of that sort but I am positive after hearing the same noises on the sound clips on this site that what we encountered was a sasquatch.

OTHER WITNESSES: 4 witness, we were trying to replace a broken chain on a dirtbike.

OTHER STORIES: Ive heard of sightings on a buddys property in Tennessee.

TIME AND CONDITIONS: It occured around 9:30/10:00, it was very dark the moon was covered by thick overhead trees and very wet it had rained the day before.

ENVIRONMENT: Very dense brush, you can barely walk through anywhere off the trail, lots of small hills.

Follow-up investigation report by BFRO Investigator Eric Lester:

I spoke to one of the witnesses at length about the incident. It occurred in deep forest outside of Elberfeld, while the group was fixing a dirtbike along a trail. A series of low, mournful moans were heard that lasted about 6-7 seconds each. The moans were very similar to the Ohio howl, but a little deeper sounding. After the moans, the forest got really quiet. Some woodknocks were heard shortly after - there were a series of three knocks, a pause, then three more. The group of friends started getting frightened from all this happening right by them in the forest. They then heard something with long, heavy strides run around them, about 30-40 yards into the forest. While the creature was running, they decided they had enough and took off out of the area.

The area was described as being an old coal mining area, with lots of creeks, deer, and thick forest cover.

About BFRO Investigator Eric Lester:

Eric Lester is a medical professional and has been investigating possible Sasquatch sightings since 2004.

BFRO expeditions include:

2005 CA Redwoods Expedition

2006 Ohio Expedition

2006 Wisconsin II Expedition

2006 West Virginia Expedition (Pocahontas County)

2007 Michigan UP, Utah, Arkansas

2008 Texas, W. Virginia, Michigan U.P.

2009 Tennessee, Wisconsin, Michigan U.P., Oklahoma

2010 private expeditions in Illinois/Iowa

++++++++++++++++++++++++
++++++++++++++++++++++++

YEAR: 1985

SEASON: Winter

MONTH: February

DATE: not sure

STATE: Indiana

COUNTY: Vermillion County

LOCATION DETAILS: Close to Newport Chemical Depot

NEAREST TOWN: Newport

NEAREST ROAD: U.S. 41

OBSERVED: In 1985, I was attending Indiana State University in Terre Haute, IN. One Saturday morning I was travelling to Chicago to visit relatives. I was driving north on U.S. 41 and I looked to my right down into a small valley (more a depression in the land than a valley) with a small stream running through it. About 100-150 yards from the highway, I saw a large dark-colored creature walking away from the road (to the east.) It was very broad and I could clearly see the creature's arms swinging. My view of the bottom part of it was obscured by tall grass and I could only see the very tops of its legs. The head never turned at all. I saw this for 4-5 seconds before the trees on the north side of the depression obscured my view. There were also trees on the south side. My immediate thought was that I didn't really see what I thought I did. I continued driving, unsure of what to do. After a few minutes, I turned around and drove back past where I had seen it (my view from this side of the highway was blocked), turned around again, drove to the bridge where the sighting occurred and stopped where I had originally seen it.

There was nothing there at this time. Not being aware of any sasquatch sightings in this area (or the Midwest in general), and not knowing what else to do, I continued on to Chicago and didn't think much more about it for many years.

ALSO NOTICED: Just prior to this sighting, maybe 5 or 10 minutes before, I had stopped and taken pictures of a red-tailed hawk that I had seen sitting on a fence along the highway. Being somewhat of an amateur photographer and naturalist, I was keeping an eye out for any other animals or picture opportunities I might come across. So at the time of the sighting, I was actively paying attention to what I was looking at (not just daydreaming, etc.)

OTHER WITNESSES: none

TIME AND CONDITIONS: Mid-morning; slightly overcast;

Follow-up investigation report:

The witness is a professional person who travels quite a bit. We hope to meet the next time I am through that area.

++++++++++++++++++++++++++
++++++++++++++++++++++++++

YEAR: 1988

SEASON: Fall

MONTH: October

STATE: Indiana

COUNTY: Vermillion County

LOCATION DETAILS: Vermillion County, Indiana ; Near Broulettes Creek, approximately 2.5 miles east of SR 63

NEAREST TOWN: Clinton

NEAREST ROAD: SR 63

OBSERVED: A friend and I had a cookout one Friday evening and kept hearing horrible screams and smelling some nasty smell. We finally went in for the night and decided to investigate on Saturday. While out on Saturday, we walked into the heavily wooded area following a power line. The whole time we were out, we felt as if we were being followed. Closer to the creek, we began to see a large, biped figure moving toward the creek. We got scared and headed back toward the road. On the way, we noticed a large footprint in the fresh mud. Later in the month, we found unusual hair on a barbed wire fence. Our zoology teacher ran tests, but couldn't determine the origin. In the fall of every year now, we go out looking and are still finding unusual things.

I'd love to help some "real" researchers seek bigfoot in this area. I'm very well-educated and can be trusted.

ALSO NOTICED: Constantly hearing noises of "diabolical screams".

OTHER WITNESSES: Cookout, then seeking out the creature

TIME AND CONDITIONS: Sounds heard every fall since

ENVIRONMENT: Heavily wooded areas near Broulettes Creek in southern Vermillion County. About 4 miles southwest of Clinton. Actually closer to Shepardsville in Vigo County.

Follow-up investigation report:

Initial contact with the witness was made on 3/28/98. He was asked to give an account of the intial report filed.

Phil and his friends went out to investigate the sounds they were hearing down by the river. That is when they saw a track going up the bank. Witness was asked to relate specifics on the footprint. Phil stated that he could make out three toes, but thought there were more. It was apparent that whatever left the track slipped on the river embankment.

It was right after the discovery of this track when the witnesses saw this thing that was walking on two legs.

The unusual hair was found about 5 - 5 1/2 feet up on a barbed wire fence. The witness said that he never did tell his teacher what he thought the hair was..

Witnesses described the sounds as a women screaming , but higher pitched and at great lengths with an abrupt cutoff. This would go on for some time and then stop only to start up again.

Phil stated that there were unusual smells in the area when the sightings occurred. He described the odors as smelling like "strong, stagnent feces."

Investigated by Steve Jackson

++++++++++++++++++++++++++
++++++++++++++++++++++++++

YEAR: 1980

SEASON: Summer

MONTH: June

STATE: Indiana

COUNTY: Vanderburgh County

LOCATION DETAILS: Wooded suburban,rural area north of city of Evansville,In. Area bordered by Evergreen Rd., Darmstadt Rd., Old State Rd., & Wortman Rd.

NEAREST TOWN: Evansville

NEAREST ROAD: Berry Drive, near U.S. Route 41

OBSERVED: I was going to feed rabbits I raised, when I heard someone walking through the brush, when I looked I saw a dark figure of what looked like a man. when I said "hello" it swayed from side to side and studied me, then began to approach me, I then could see very deep eye sockets oddly shaped, and it was covered in black hair. I turned and ran to the house, no sounds were heard. I was twenty feet or less from the creature when I saw it.

ALSO NOTICED: In a nearby woods my friend was riding his mini bike. He had stopped to take a break when this creature charged through the woods into the clearing where he was stopped. It was sweating, with a beet red face and winded. He started his bike, startling the creature. He took off to home and he also told me of seeing strange feces in a nearby abandoned barn during the same year.

OTHER WITNESSES: Just myself. I also have a friend who saw the same creature about the same time that year in a nearby woods

OTHER STORIES: See above

TIME AND CONDITIONS: Between 5:00-6:00am,Very quiet,still, just before sunrise.Warm & clear.

ENVIRONMENT: Suburban yards between two lakes, also nearby wooded uninhabited {by people} areas. Sighting area connects to large woods ,ash, hickory, oak, trees etc. Said to be small cave in center of area. My actual sighting spot was a

vacant lot behind home, very dense thicket, creature walked through. Also railroad right of way in this area

Ditches, valleys, hilly.

++++++++++++++++++++++++
++++++++++++++++++++++++

YEAR: 1978

SEASON: Summer

STATE: Indiana

COUNTY: Sullivan County

LOCATION DETAILS: 1/2 mile north of the Glendora Road Lake Sullivan causeway, near Sullivan, Sullivan County, Indiana, USA.

OBSERVED: In late summer, 1978, (specific date is being researched), the Sheriff telephoned me regarding a previous night report from a youth who related he and his girlfriend saw a sasquatch. A deputy responded to the called in report about 11 PM the evening before and took a short report but developed no further information. As much for curiosity as anything I located the reporter of the incident and secured a tape recorded statement. He related he and his girlfriend has returned to her home located on the east side of lake Sullivan at about 11 o'clock PM and were talking, she sitting on the car right fender and he embracing her. The car was parked facing west towards the rear yard, security light, low stock fence and bean field behind it. At the same time both subjects observe a large biped animal type creature appear under the overhanging outside light about 100 feet away. It initially was crouched over, then stood up, stared at the two observers, hesitated for a few seconds; then stepped over the fence and disappeared into the soybean field to the west. Horses in a field next to the lake to the west were heard spooked and running. The creature appeared to be eight feet tall, covered with hair, eyes were distinctive. Next morning search of the bean field determined impressions in the dirt but were not distinctive - it had rained during the night. There was no further supporting evidence other than the girl's mother states the couple were very shook when they ran into the house to call the Sheriff. I interviewed the youth 10 years later on a whim to determine if there was a different story or to per chance get a denial of the original report. To my surprise he retold the account in precise detail as compared to the original. A separate report initially seeming to have no connection was reported to the Sheriff the morning after from Hymera, a small town eight or nine miles northeast of the sighting location regarding a rabbit hutch had been torn completely apart by perhaps a vandal. All rabbits were gone. The rabbit hutch site is directly through the country from the alleged sasquatch sighting connected with coal stripped lands and creek bottoms that feed Lake Sullivan

++++++++++++++++++++++++
++++++++++++++++++++++++

YEAR: 2000

SEASON: Fall

MONTH: November

DATE: 11/18/00

STATE: Indiana

COUNTY: St. Joseph County

LOCATION DETAILS: My house is about 7 miles away from Notre Dame

NEAREST TOWN: South Bend

NEAREST ROAD: Kern

OBSERVED: Me and my friend Mike were hiking in some woods not too far from my house in northern Indiana when we heard a noise. We (being 13 year old kids) decided to check it out. We kept finding peices of wood around some foot prints. Then I heard another noise and when I looked over my shoulder there was a hairy ape-like being about 6'8" no more than twenty yards away. Our instincts were to run, so we did. When we returned to my house my mom did not believe us. That was my run in with big foot.

ALSO NOTICED: nothing out of the ordinary.

OTHER WITNESSES: my friend Mike.

OTHER STORIES: I have heard about the Sister Lakes Monster, (nearby area) and saw a book at the library with an article about it.

TIME AND CONDITIONS: During the day at about 5:00-7:00

am low light at dawn damp and foggy.

ENVIRONMENT: It was in a forest neer a small creek.

Follow-up investigation report:

I have spoken with the mother of this witness twice, and the witness once. We will keep in touch, and he is going to find the title and author of the book he mentioned about the Sister Lakes Monster.

++++++++++++++++++++++++
++++++++++++++++++++++++

YEAR: 2000

SEASON: Fall

MONTH: November

DATE: 11/18/00

STATE: Indiana

COUNTY: St. Joseph County

LOCATION DETAILS: My house is about 7 miles away from Notre Dame

NEAREST TOWN: South Bend

NEAREST ROAD: Kern

OBSERVED: Me and my friend Mike were hiking in some woods not too far from my house in northern Indiana when we heard a noise. We (being 13 year old kids) decided to check it out. We kept finding peices of wood around some foot prints. Then I heard another noise and when I looked over my shoulder there was a hairy ape-like being about 6'8" no more than twenty yards away. Our instincts were to run, so we did. When we returned to my house my mom did not believe us. That was my run in with big foot.

ALSO NOTICED: nothing out of the ordinary.

OTHER WITNESSES: my friend Mike.

OTHER STORIES: I have heard about the Sister Lakes Monster, (nearby area) and saw a book at the library with an article about it.

TIME AND CONDITIONS: During the day at about 5:00-7:00

am low light at dawn damp and foggy.

ENVIRONMENT: It was in a forest neer a small creek.

Follow-up investigation report:

I have spoken with the mother of this witness twice, and the witness once. We will keep in touch, and he is going to find the title and author of the book he mentioned about the Sister Lakes Monster.

+++++++++++++++++++++++++++
+++++++++++++++++++++++++++

YEAR: 1968

SEASON: Summer

MONTH: July

STATE: Indiana

COUNTY: Spencer County

LOCATION DETAILS: As you travel south on highway 131 from Gentryville, you will hit a curve that turns east from an intersetion with a highway that comes from boonville, IN. At a sharp curve back to the south towards Chrisney, IN, you turn off to the east and follow a one lane country road east to an T intersection with a another country road. You will be about 1/4 mile off 131. Then you turn left on this gravel road and head north. You will see the old barn and farmhouse just to your right. Today a cell tower stands on the property which makes it easy to spot and identify.

NEAREST TOWN: Chrisney

NEAREST ROAD: country gravel road or Highway 131

OBSERVED: I was just a young boy then staying at my grandfather's farm in the summer time. I slept up in the attic at night with my uncle Donnie who was helping my grandfather with his dairy farm. We had just laid down when my uncle Donnie got right back up and headed downstairs. I was on my stomach looking East out the window. (We had two windows in the attic for ventilation. One was on the west and the other was on the east end.)

We had a small brooder house just east of the farm house. I also was looking out over the kitchen roof which was actually an add on to an old house porch. Suddenly a "man" walked out in the open from just north of the brooder shed as a truck pulled into the driveway. At the same time a truck pulled into the drive way.

The driveway came up the hill (the one lane gravel road ran north to south) to the east and then curved toward the north (to the kitchen area). The headlights glanced or lit up the creature. Three things I noticed. First the hair, face, or body did not shine or reflect any light. The creature was very dark. This could be one reason big foot is difficult to spot or see in wooded areas. Secondly, I realized it did not look like a man because the arms were abnormally long. Just like a man the arms were swinging, but closer to the knees than the ordinary person. Height appeared close to that of a normal adult male. I would say under six foot tall but not under five foot tall. The head was fairly rounded.

The creature dashed off to the west, and quite frankly I was petrified with fear. My uncle Donnied liked to scare me with "Boogie Man" stories, and I had just seen my first real live one.

A little later the truck left and my uncle Donnie came up. I told him I had seen somebody outside by the shed. He went back downstairs, told my grandfather, and they grabbed their shotguns and searched around the house, and then drove the old ford pick-up up and down the road several times with flash lights. They did not see or find anyone.

ALSO NOTICED: Despite all the disturbances and changes in the area, some interesting things have occurred since then. Coyotes returned in large numbers, bobcats made a come back and are quite common, my grandfather claims to have seen a small black bear in the intervening years on the farm (about 16 years ago), turkey, and other kinds of wildlife. Lincoln State Park is to the north about ten miles of my grandfather's farm and a strip of woods still covers a lot of the area between the farm and the state park.

I suspect this big foot was a young animal searching for a place to live. Minining had just started, farms were becoming more intensive, and other changes were taking place. The return of predators might have also caused the animal to relocate. (A black bear might have been more than a match for young big foot). What still strikes me was the non relective quality of the fur. If what I saw was truly big foot, this is an animal that moves in the shadows of deep forest, and at night. Day light observations must be abnormal in this part of the world. The other thing that struck me was the animal was used to humans, but reacted quickly when head lights hit it. It did not freeze like so many other animals will such as deer. I suspect it doesn't like large groups of people. That the best or most likely encounters are when the animal doesn't feel exposed or when only one or two other humans are around. This would fit the behavior of an animal who had mostly solitary and scattered territories.

I think to catch a big foot - you must work alone or in pairs (too many people drives them off), be willing to set up blinds where you lay down (standing up may provoke a natural flight response in young animals and may allow the animals to see you), your best chances of spotting one is at night under thick canopies or moonless nights, and the animal can not only smell, see and hear you, he has above average intelligence. Or at least he is smart enough to bolt instead of staring into the headlights. Someone once told if a turkey could smell as well as he could see and hear, you would never manage to shoot one. Throw a little simian intelligence and you have a real job of ever finding or catching or shooting one. One other thing that really struck me. He didn't follow a natural ridge or water way like most animals do. He was cutting across terrain featuress. The one advantage perhaps an upright ape has over other creatures. He doesn't follow or trek across the easiest routes or is funneled by terrain. Something I could speculate about again.

OTHER WITNESSES: No. No one else saw it.

OTHER STORIES: No. Except for the occassional story in the Evansville Courier and Press.

TIME AND CONDITIONS: The incident occurred right after sun down.

Somewhere between 8 and 9 PM since we usually went to bed by then.

ENVIRONMENT: The environment was in the open, next to farm land in a three crop rotation with dairy cattle. The house sits on top of a small hill. The creature was coming up from a field and drainage ditch and from the vicinity of a small farm pond. At the time the east side of the farm was heavily wooded and far fewer houses existed in the area. Principally three farmers farmed the land in and around this area to the south and east of the farm. On the otheside of the water shed or drainage at the time was forest wood lots and abandon farm fields. The land is hilly and the Ohio River is only several miles south of there. In the years since this incident, the land east and south of the farm has been developed. More houses and people now exist along the roads, a power plant was built just south of Chrisney near Rockport Indiana. Also the land east was heavily stripped mined for coal.

Follow-up investigation report by BFRO Investigator Tony Gerard:

I spoke with the witness by phone. He was about eight or nine years old at the time of his sighting. The creature was in view from the headlights for about three seconds. It "turned and looked toward the lights, then took off running". The gait was described as "odd, long strides with its long arms swinging by its sides". The arms seemed to reach below knee length. He described the body build as on the thin side and it did not seem to have a distinct neck. He had no real impression about the length of the body hair. He stressed how unreflective the creature was, that even it's face did not seem to reflect much light.

++++++++++++++++++++++++
++++++++++++++++++++++++

YEAR: 1973

SEASON: Summer

STATE: Indiana

COUNTY: Shelby County

LOCATION DETAILS: Smithland road in southern Shelby county. It was at a ninty degree turn in the road where I think it changed from East to North direction.

NEAREST TOWN: Flat Rock

NEAREST ROAD: Smithland Road

OBSERVED: I was driving home from a girlfriends house in the early hours of the morning on a late summer night. Patches of ground fog were scattered about over the fields. As I drove up to a ninety degree turn in the road my headlight shown on a figure in a bean field that boardered a corn field. The headlights caused the eyes to shine red and I could make out the shape of a large figure. It was standing in an area of the fog and its head and upper body were above the fog. It was as tall as the corn in the field next to it. Since the figure was next to the corn field (standing in the beans) it gave frame of referance for height. It was as tall as the corn which in Indiana at that time of the year is at least seven to eight feet tall. I could not make out facial features or hair because of the fog and the fact that it was at least twenty yards into the field. I turned at the curve and punched the gas. My hair was standing up on my arms and head and it really scared me. I was by myself so I was not about to hang around. I did not tell anyone as I thought that I would be ridiculed.

ALSO NOTICED: None

OTHER WITNESSES: None

OTHER STORIES: None.

TIME AND CONDITIONS: 1:00 - 3:00 am Moonlit night. Small patches of fog that where approximately 5-6 feet above the ground.

ENVIRONMENT: The area is farmland with rolling hills puncutated with areas of woods and small streams.

Follow-up investigation report by BFRO Investigator Stan Courtney:

I spoke with the witness by phone. Since he did not stay longer and watch the animal he was unable to see any features or movement.

About BFRO Investigator Stan Courtney:

Stan Courtney has a special interest in bird and wildlife audio recording. Stan has attended numerous BFRO Expeditions.

Stan Courtney can be reached at illinois@bfro.net

~ 177 ~

++++++++++++++++++++++++
++++++++++++++++++++++++

YEAR: 1992

SEASON: Fall

MONTH: October

DATE: 20

STATE: Indiana

COUNTY: Scott County

LOCATION DETAILS: take I-65 to Austin IN go west on 256 turn left on Boatman Rd turn right on York street, fallow the road till it turns into gravel and then dead ends at the river in the middle of a field. It is the deep in the woods on the right

NEAREST TOWN: Austin

NEAREST ROAD: Boatman Rd , I-65

OBSERVED: It was bow season in October 1992, around 5:30 in the evening. I had just climed out of my deer stand and heard movement about 20 meters in front of my stand. So I decided to just sit at the base of the tree to see if if the deer would come out of the brush, and hopefuly get a shot at it. After waiting for about 10 minutes it got pretty dark in the woods. Too dark to take a shot, so instead of spooking the deer, I decided to wait for it to move out so that I might have another chance at it the next day. It was making alot of noise, kind of unusual for deer. I must have set there for another 20 minutes when I heard this loud high pitch scream, or howl. It lasted for about 3-4 seconds and it was very close to my left. Needless to say, it scared the crap out of me! I have never heard anything like that before! I grew up around the woods and I have been hunting for many years, and never heard that sound before or since. I have been in the military for 16 years and spent my share of many different types of wilderness, and this sound that I heard I will never forget.

ALSO NOTICED: nothing and I hunted these woods for years

OTHER WITNESSES: no

OTHER STORIES: no

TIME AND CONDITIONS: between 6:00 - 6:30 pm. dark in the woods, but you could just see the sky through the tree tops. and it was around 55degrees

ENVIRONMENT: the area is called the Austin Bottoms, because it is always flooded, 1000's of acres of woods, alot of farm land. many branches of the muscatak river run through it. lots of game

---

Follow-up investigation report by BFRO Investigator Eric Lester:

I spoke to the witness, an Iraq war veteran, over the phone about what he heard that evening. The scream/howl was so close to him that he jumped up and ran while the howl was still going on, and he didn't stop until he was out of the area and back to the road, about a half-mile away. When he initially started running, the scream/howl ceased. It was described as rising in pitch, extremely loud, and similar to the howls found on the BFRO website.

The area is described as being extensive forest interspersed with many swamps. The Muscatatuck River runs through this area.

He hasn't been back to those woods since.

++++++++++++++++++++++++
++++++++++++++++++++++++

YEAR: 2006

SEASON: Summer

MONTH: August

STATE: Indiana

COUNTY: Putnam County

LOCATION DETAILS: It was at Glen Flint Lake. The area where the sounds came from were at the end of the cove we were in straight out from the boat ramp to the left before you get into the main lake area.

NEAREST TOWN: greencastle

NEAREST ROAD: US 36

OBSERVED: My girlfriend and I were fishing on Glen Flint Lake in Putnam Co. IN. I would say it was about 1 am. We heard a coyote, or coyotes barking and making yipping noises, then we heard several loud grunt like noises, almost moan like, very loud similar to what a bear sounds like, then a coyote let out a scream like sound, like it was getting hurt. Then it got quiet. I pulled up our anchors, and left. I have camped all over the United States, I've camped in remote parts of Canada. I have never heard anything like that. I was actually a little scared. Whatever it was it was big. To this day I still believe it was a bigfoot. If I had to guess, the coyote didn't make it.

ALSO NOTICED: I have been there since, and heard a grunt similar to what we heard that night, but it wasn't several, just one loud one, almost moan like.

OTHER WITNESSES: My girlfriend Tabatha

OTHER STORIES: I just read about sounds that someone hears who posted on this website, it sounds like they live close to where heard the sounds.

TIME AND CONDITIONS: It was warm, around 1 am, calm. a nice summer night for fishing

ENVIRONMENT: A public fishing and recreational lake.

Follow-up investigation report by BFRO Investigator Eric Lester:

I spoke to the witness by phone about what he heard that night. He decided to submit his experience after reading Report #25237, which is very near this area. The sounds he heard occurred about 500 yards away from them. The coyotes were yipping and barking for 30-40 seconds before the first grunt/moans were heard. They were described as loud, and lasting about a half-second each. These grunt/moans ended after about 15-30 seconds, and nothing else was heard.

The witness stated that he has much camping experience and some encounters with bears. He seems to think the grunt/moans are similar, but not quite identical, to what a bear can make.

Glenn Flint lake is a recreational lake built in the 1970s, and is in the same area as Van Bibber lake. This area also contains many streams and creeks.

++++++++++++++++++++++++
++++++++++++++++++++++++

YEAR: 2008

SEASON: Winter

MONTH: December

STATE: Indiana

COUNTY: Putnam County

LOCATION DETAILS: One hour west of Indianapolis on US 36, before US 231. Farmland, fields, and dense woods.

NEAREST TOWN: Greencastle, IN

NEAREST ROAD: US 36

OBSERVED: I have reported before on this website. I live in a rural area in west-central Indiana. I have heard strange primate-like calls coming from the woods behind my residence for about eighteen months. About two weeks ago, I heard powerful whooping calls coming from our woods as I was changing a tire on my car at about 10PM. These calls were followed by wood knocks. The calls sounded like those I have heard on this site. I have been an amateur bigfoot researcher for many years, and am convinced that one, possibly more, are inhabiting the woods behind my house.

OTHER WITNESSES: None.

OTHER STORIES: Other incidents experienced by myself and my father in the same general area.

TIME AND CONDITIONS: 10PM, dark, clear weather, cold.

ENVIRONMENT: Heavily forested area about 180 acres total. Hills, a creek system, and fields surrounding.

Follow-up investigation report by BFRO Investigator Eric Lester:

I interviewed the witness by phone about what he, and on some occasions his father, have heard near their home. They started hearing these whoops and howls in the summer of 2006. These sounds always occur at night; sometimes starting after dusk and lasting until 3-4 am, or starting around midnight, but lasting only an hour or so. Yet there is no set pattern on when these sounds will be heard. In other words, it cannot be predicted when they will occur.

The howl is described as starting very low and drawn out, and rising in pitch to a high scream, then tapering off (about 6-10 seconds in length). If a response occurs, it is the same type of howl, and usually comes from the opposite side of the woodland behind the house. This happens within a minute of the first howl. The whoops (such as the ones heard while changing his car tire) are quick and high-pitched, usually 3 or 4 at a time.

During July of 2008 the witness and a friend camped in the forest behind his house. Prior to settling in for the night, they made their own 'whoop' sounds and wood knocks. About 12:30 AM they got 3-4 whoops in reply, which sounded about 40 yards away from their camp. These were described as coming from something with 'a lot of lung capacity'. After 10 seconds the whoops occurred again, along with some movement in the brush from the general direction the noises were coming from. Nothing else happened that night.

From another submitted report by the witness, he stated - 'What struck me initially was the power behind these calls; it would take an animal with tremendous lung capacity to make them carry like they do.' This other report was not published.

The area near the home contains many pockets of forests among the farm fields. There are also many small waterways connecting these pockets, and the area is heavily populated with deer and turkey.

About BFRO Investigator Eric Lester:

Eric Lester is a medical professional and has been investigating possible Sasquatch sightings since 2004.

BFRO expeditions include:

2005 CA Redwoods Expedition

2006 Ohio Expedition

2006 Wisconsin II Expedition

2006 West Virginia Expedition (Pocahontas County)

2007 Michigan UP, Utah, Arkansas

2008 Texas, W. Virginia, Michigan U.P.

2009 Tennessee, Wisconsin, Michigan U.P., Oklahoma

2010 private expeditions in Illinois/Iowa

++++++++++++++++++++++++
++++++++++++++++++++++++

YEAR: 1953-1976

SEASON: Summer

STATE: Indiana

COUNTY: Pulaski County

LOCATION DETAILS: All the sightings were near the Dentwood cabins, right off River Road, just outside of Winamac.

NEAREST TOWN: Winamac

NEAREST ROAD: River Road

OBSERVED: These are my husband and his family's accounts.

The first sighting was in 1953. The Tippecanoe River runs beside a cluster of cottages near the little town of Winamac, in northern Indiana. A long suspension bridge, spans the river, and was the only means of getting to the five cottages, other than by boat. At each end, flood lights were mounted, and at night, each lighted the bridge half way across. The bank of the river was rather steep. Behind the group of cottages, the woods was thick and dense. Children were not allowed to wander there, and play was confined largely to the area right around the cabins.

Two brothers, each with large families, had come for the weekend to enjoy fishing, boating, and just generally, being together. My husband, one of the children, was then about five. On this particular day, a group of little girls, his sisters and cousins, were playing between two of the cabins. The grownups, and some of the children, were inside.

About the middle of the day, terrified screams came from the group of girls. When the adults, hearing the screams, ran outside, they found the girls crying hysterically, and pointing towards the woods. A huge creature, about seven feet tall, walking on two legs, covered from head to toe, with long dark hair, was walking the other way, towards the treeline. He was walking upright, on two feet. One child was saying that it had touched her, others were saying that it had just one eye.

A rifle was was brought out from the cabin, and one man, my husband's father took aim. He had the beast in his site, held the gun for a few seconds, and then lowered it. Cries of, "Shoot him," could not persuade him to fire at the beast. When asked later, why he didn't shoot, he said, "I was afraid that I might be shooting a man." One of his children, now, my husband, came running outside with the adults and saw the creature's back as he walked towards, and into the woods.

Ten years later, in 1963, my husband,(then fifteen), and his family, were once again spending the week in one of the cottages. Late one night, after all the others had gone to bed, he was down on one of the docks. Loving the river at night, he often walked it's banks. On this particular night, he began to hear a movement coming from behind the cabins at the edge of the woods. He walked up the bank, and went around to the back of the cabin to see what was making such a ruckus. There, in the darkness, coming out of the woods slowly, was the same creature that had frightened the family ten years ago. He ran back to the front porch, hurried into the house, and woke up his father. The two of them, without waking the others, went onto the screened porch, which was in front, towards the river, and in the darkness, with a 22 rifle and only the bridge lights shining, they waited.

They began to smell a terrible stench, like that of rotten meat, getting worse as the creature neared them. This cabin was right next to the bridge. They knew that if the creature was going to cross it, he would walk right by them, about ten feet away, and pass under the light. Inside the cabin, the family dog was beginning to whine. They stood quietly, and soon the creature was beside them, about ten feet away. He seemed to be intent on going for the bridge. As he passed the porch, he turned and looked towards the teen and his father, but seemed not to see them, and didn't regard them in any way, nor, the dog, which was now howling, and tearing at the door.

The creature made his way to the bridge. As he passed under the floodlight on the cabin side, he was clearly seen, as huge, covered with dark hair, and walking upright like a man. The smell was horrible. When he got to the bridge, he went down on all fours and began to cross it using his arms and hands, as well as his feet. When he passed the half way mark on the bridge, and neared the other side of the bridge, the dog was let out on the porch and crashed through the screen door, running after the creature. It raced across the bridge. By this time, the creature had crossed the bridge and was in the darkness beyond. The teen and his father could only guess what was happening by the sounds they heard - growling, fighting noises, and then whimpering and yelping. The dog came back to the cabin covered with cuts and blood.

The only other time that anyone said that the creature had been seen was an account told to me by my husband's neice, who often stayed in the same cabin as a child. She said she woke up one morning and saw a large, yellow, hairy creature looking through the door. He left quickly. This was about 1976.

OTHER WITNESSES: The first incident was witnessed by, maybe 8 or 9 people. Prior to the incident, the children were playing in the yard. The adults were in the house. In the second incident, there were only two witnesses. One had been at the river bank, walking, the other was asleep in the cabin. In the last incident, there was only one witness and she had been sleeping.

OTHER STORIES: No

TIME AND CONDITIONS: The first sighting was in the middle of the day, the next, 11 or 12 at night and the last early in the morning.

ENVIRONMENT: There was a thick, dense woods behind the cabins. It was in a rural farming area, right beside the Tippecanoe River. A suspension bridge is the only access to the cabins, but they are located near the road, on the other side of the river.

Follow-up investigation report by BFRO Investigator Stan Courtney:

++++++++++++++++++++++++++
++++++++++++++++++++++++

EAR: 1988

SEASON: Winter

STATE: Indiana

COUNTY: Pulaski County

LOCATION DETAILS: It is at the old Oaks Campground, the county line is less than a mile from there where it meets Jasper County.

NEAREST TOWN: Francesville

NEAREST ROAD: 114

OBSERVED: My friends and I were going for a walk and we came across a bare footprint. It wouldn't have seemed odd, but it was in the melting snow and it was the only one we could see. We ran to my parents and told them what we found and we brought a tape measure and took a polaroid picture of it. It only measured 13-14 inches long which is not very long at all but it did seem wide. I don't know how significant this information is but you never know. It was a long time ago anywhere from 10-15 years ago...I guessed on the year, I think I was anywhere from 8-10 years old.

ALSO NOTICED: It was near the trailer park where people lived...and there was no one living in the campers that were there in the winter. The footprint was headed back towards the woods.

OTHER WITNESSES: I can't remember who all were there but there were at least four of us.

OTHER STORIES: No, like I said I don't think this is too significant because we never saw anything but the footprint. I do still have the photo.

TIME AND CONDITIONS: We found it in the morning...and it was pretty cold and dreary. Footprint was melting and icy.

ENVIRONMENT: The place is surrounded by woods and fields and the highway is nearby

Follow-up investigation report by BFRO Investigator Stan Courtney:

I spoke with the witness phone. She sent me photos of the track, however they did not show enough detail to be included with this report.

++++++++++++++++++++++++++
++++++++++++++++++++++++++

YEAR: 1978-1990

SEASON: Unknown

STATE: Indiana

COUNTY: Perry County

LOCATION DETAILS: Hoosier National Forest, Patoka Lake State Recreation Area, Birdseye, IN, Bristow, IN, English, IN, Leopold, IN, intersections of State Highway 37 and Interstate 64

OBSERVED: My name is JET. Beginning around 1978 or thereabouts, the area in which I (then) lived began to report various sightings of what could only be construed as a Bigfoot Sasquatch). I live in southern Indiana, in Crawford County. At that time, I lived in nearby, bordering Perry County. The name of the town of the sighting was Bristow, somewhat near the port of Tell City, Indiana (Ohio River). A young couple sighted one while driving at night and the local newspaper of that time (Hard Times - "The Old Man Was Right") carried the story. Supposedly, samples of hair and plaster casts were taken by scientists at IU Bloomington Campus (headquarters). Nothing was ever said of what was discovered. Over the next 15 years, I have personally heard fascinating, first-hand accounts from more than 3 people (each from very different backgrounds, ages and from many different places of residence)of Bigfoot sightings since that time. An acquaitance I met at church 20 years ago told of playing hide and seek in a corn field at night and running into one. The story is chilling, believe me. Another gentlemen I used to work with recalled seeing it in broad daylight on his way to a cabin in the woods. My own father saw the creature near Patoka Lake, Indiana (Crawford, Orange, Dubuois counties) We were squirrel hunting near the lake and he was in deep cover. While stalking a squirrel, he saw what he took to be legs standing across a slogh, against a tree. Now, my father taught me to never point a gun, loaded or unloaded, at anything not intended to be killed and eaten. He would not so much as point a telescopic scope (while mounted to a rifle) at a man walking 300 yards away just to identify him. He told me that the creature began to walk away, on two feet, and he IMMEDIATELY brought up his gun (a measly .22 lever-action rifle) and trained his attention upon it. Before he could make a definte identification, the creature noticed he was being watched and strode towards the lake. When I found me Dad later that morning, he was jogging (which was indeed VERY rare) and passed me up on the way to the check-out station. He couldn't even relate the story to me until we were 10 miles away. He told me to take a .308 rifle back to that area if I even went back there - even to squirrel hunt. This man is also an ordained minister and long time memeber and pastor of a church. Furthermore, one needs not describe how one knows when one's own father is telling you the truth: let alone in such detail. Anyway, I hope you at this website are doing all of the most serious research that you can on this creature. People should know that it IS REAL AND DOES EXIST. Whether or not it means harm, who knows? I don't necessarily think so. According to my own research and documentation of the stories that I have mentioned above, all of these sightings were within the geographical boudaries of the Hoosier National Forest area. Need I say more? Look all of this up on the web:

I've never witnessed hair or tracks, but then again, I never went looking. Now don't get me wrong, I hunt just about everything that moves - deer, squirrels, foxes, crows, varmints, etc. I have no qualms about being in and around the woods. I am very familiar with about a 10 mile radius from my home - and believe me, I know what goes on most of the time. I also know that the people who have told me their stories are the same way. These people don't see a deer and think that it's a goat and so forth. I don't want to hunt this thing, I want to get someone to prove that it's there. I have heard very strange sounds from the dark valley behind my father's house, but I believe that we finally agreed (after years of debate) that what we were hearing was probably a bobcat (lynx) or something very similar. Stories of what are commonly called "sugar bears" abounded in the English area years ago. Many reports of panthers also exist. Another area of acitivity is the Hemlock Cliffs area. This is a long forgotten cliff and valley layout that is deep into the middle of nowhere. Many strange things have been sighted and reported thereabouts.

ALSO NOTICED: Not too many reports of these types of things. Again, I don't know anyone who went back looking for something afterwards

OTHER WITNESSES: Driving down a road at night, walking with their children and admiring the construction of a new, small lake, driving a truck through a field and proceeding to a cabin, squirrel hunting (stalkinig, stealthily).

ENVIRONMENT: Check near the area of Louisville, Kentucky. We are due west of Louisville, about 45 miles out towards St. Louis. Look on the map at where State Highway 37 runs North to Indianapolis. One was in a cornfield, one was walking around a small lake, one was in a hay-field near a forest, and the other was near Patoka Lake in southern IN (8,880 acres at maximum water pool.

++++++++++++++++++++++++
++++++++++++++++++++++++

YEAR: 1978-1990

SEASON: Unknown

STATE: Indiana

COUNTY: Perry County

LOCATION DETAILS: Hoosier National Forest, Patoka Lake State Recreation Area, Birdseye, IN, Bristow, IN, English,

IN, Leopold, IN, intersections of State Highway 37 and Interstate 64

OBSERVED: My name is JET. Beginning around 1978 or thereabouts, the area in which I (then) lived began to report various sightings of what could only be construed as a Bigfoot Sasquatch). I live in southern Indiana, in Crawford County. At that time, I lived in nearby, bordering Perry County. The name of the town of the sighting was Bristow, somewhat near the port of Tell City, Indiana (Ohio River). A young couple sighted one while driving at night and the local newspaper of that time (Hard Times - "The Old Man Was Right") carried the story. Supposedly, samples of hair and plaster casts were taken by scientists at IU Bloomington Campus (headquarters). Nothing was ever said of what was discovered. Over the next 15 years, I have personally heard fascinating, first-hand accounts from more than 3 people (each from very different backgrounds, ages and from many different places of residence)of Bigfoot sightings since that time. An acquaitance I met at church 20 years ago told of playing hide and seek in a corn field at night and running into one. The story is chilling, believe me. Another gentlemen I used to work with recalled seeing it in broad daylight on his way to a cabin in the woods. My own father saw the creature near Patoka Lake, Indiana (Crawford, Orange, Dubuois counties) We were squirrel hunting near the lake and he was in deep cover. While stalking a squirrel, he saw what he took to be legs standing across a slogh, against a tree. Now, my father taught me to never point a gun, loaded or unloaded, at anything not intended to be killed and eaten. He would not so much as point a telescopic scope (while mounted to a rifle) at a man walking 300 yards away just to identify him. He told me that the creature began to walk away, on two feet, and he IMMEDIATELY brought up his gun (a measly .22 lever-action rifle) and trained his attention upon it. Before he could make a definte identification, the creature noticed he was being watched and strode towards the lake. When I found me Dad later that morning, he was jogging (which was indeed VERY rare) and passed me up on the way to the check-out station. He couldn't even relate the story to me until we were 10 miles away. He told me to take a .308 rifle back to that area if I even went back there - even to squirrel hunt. This man is also an ordained minister and long time memeber and pastor of a church. Furthermore, one needs not describe how one knows when one's own father is telling you the truth: let alone in such detail. Anyway, I hope you at this website are doing all of the most serious research that you can on this creature. People should know that it IS

REAL AND DOES EXIST. Whether or not it means harm, who knows? I don't necessarily think so. According to my own research and documentation of the stories that I have mentioned above, all of these sightings were within the geographical boudaries of the Hoosier National Forest area. Need I say more? Look all of this up on the web:

I've never witnessed hair or tracks, but then again, I never went looking. Now don't get me wrong, I hunt just about everything that moves - deer, squirrels, foxes, crows, varmints, etc. I have no qualms about being in and around the woods. I am very familiar with about a 10 mile radius from my home - and believe me, I know what goes on most of the time. I also know that the people who have told me their stories are the same way. These people don't see a deer and think that it's a goat and so forth. I don't want to hunt this thing, I want to get someone to prove that it's there. I have heard very strange sounds from the dark valley behind my father's house, but I believe that we finally agreed (after years of debate) that what we were hearing was probably a bobcat (lynx) or something very similar. Stories of what are commonly called "sugar bears" abounded in the English area years ago. Many reports of panthers also exist. Another area of acitivity is the Hemlock Cliffs area. This is a long forgotten cliff and valley layout that is deep into the middle of nowhere. Many strange things have been sighted and reported thereabouts.

ALSO NOTICED: Not too many reports of these types of things. Again, I don't know anyone who went back looking for something afterwards

OTHER WITNESSES: Driving down a road at night, walking with their children and admiring the construction of a new, small lake, driving a truck through a field and proceeding to a cabin, squirrel hunting (stalkinig, stealthily).

ENVIRONMENT: Check near the area of Louisville, Kentucky. We are due west of Louisville, about 45 miles out towards St. Louis. Look on the map at where State Highway 37 runs North to Indianapolis. One was in a cornfield, one was walking around a small lake, one was in a hay-field near a forest, and the other was near Patoka Lake in southern IN (8,880 acres at maximum water pool).

+++

YEAR: 1981

SEASON: Summer

MONTH: June

STATE: Indiana

COUNTY: Perry County

LOCATION DETAILS: Lake Tipsaw

NEAREST TOWN: Leopolde

NEAREST ROAD: Hwy 37

OBSERVED: My wife and I had a sighting in June 1981 at Tipsaw Lake in Perry County. We had taken a picnic lunch with us to the park and beach area, only to find it under construction and closed. We did enter the beach area by way of the boat ramp area where we parked our car. The area was closed that day and not a single person or fisherman was on the lake. We had two small boys who immediately began to swim at the end of beach area next to the boat ramp. My wife and I were under a shade tree about 15 yards off the beach watch the boys swim. My wife was putting the picnic together while I was relaxing and looking across the lake. I could see a tall man moving between the popular trees very strangely high on the hill side near the dam to our right. The boys were having a great time, playing in the water making much noise. The tall dark man behind to move down the hillside to the edge of the water looking toward our boys in the water. The strange looking person or bigfoot began pacing back and forth looking in the direction of our boys. I continued watching but could not believe what I thought I was seeing. I then stop my wife and ask her to look toward the dam and tell me what she was seeing. My wife and I observed this being for another minute or two when he stopped pacing and began moving up the dam in a diagonal manner some 1200 yards away. My wife said we our leaving, get the boys and lets get out of here. We then headed back for Crawford Co. and have never forgotten that day. We returned to Tipsaw for the first time this past summer and looked at the spot that invoked all that fear some 27 years ago.

OTHER WITNESSES: None, My wife and I

OTHER STORIES: Yes, just that some drivers in the area along I-64 junction with Hwy 37 had made similar sighting at the time.

TIME AND CONDITIONS: About noon, excellent light, if only I would of had a camera, what a picture that would have been.

ENVIRONMENT: Lake with secondary growth woodland around

Follow-up investigation report by BFRO Investigator Harold Benny:

I talked with the witness and found him to be credible. Although the sighting was several years ago, it was still fresh in his mind. We revised the sighting distance to about 300 yards. Total observation time was about eight minutes. He watched the bipedal animal first descend the hill and then pace up and down the water's edge while watching the boys swim. The witness was adamant that the tall black creature could not have been a bear or man.

About BFRO Investigator Harold Benny:

Harold Benny holds two degrees in Zoology and was a Biology teacher for several years. He participated in the following BFRO expeditions: Michigan UP 07, Arkansas I 07, Arkansas II (Oklahoma) 07, N. Florida 08, Missouri 08, Tennessee 09, and Ohio 09.

++++++++++++++++++++++++
++++++++++++++++++++++++

YEAR: 1997

SEASON: Fall

MONTH: October

DATE: 10/01/97

STATE: Indiana

COUNTY: Parke County

LOCATION DETAILS: About 3 to 4 miles south of the Turkey Run State Park & School on Hwy 41.

NEAREST TOWN: Marshal

NEAREST ROAD: Hwy 41

OBSERVED: We were then living in Ohio. My husband had photography work to do and was being transferred to Indiana.

In the fall of 1997 while driving south on Hwy 41 near Turkey Run State Park in IN, my husband and I saw what we affectionately call "A something".

It was a misty night and we were headed for Terra Haute. I was driving my Ford Aerostar Van with my husband in the passenger seat (I was glad he was on that side) we also had our two year old son in the back seat. "It" was walking in the gully. "It" stood taller than the van so to speculate on the height I would say over 6" easily.

The body was black and hairy although the face was huge and long in the jaw shaped like a knight of a chess piece but not like a horse! My husband describes the face more like a lions shape.

We passed it and continued our trip - and later even moved here with our child going to Turkey Run School. We travel the same road often from Turkey Run to Rockville and have never seen "It" again - YET!

ALSO NOTICED: Nope, except the hairs on the backs of our necks were raised. We were asking each other "Do you see that!" "Did you see that!" Then we described to each other what we saw.

OTHER WITNESSES: Myself

Husband talking in the van

Our child in the back seat was asleep

OTHER STORIES: My brother claims to have seen one - but he travels alot very modestly. He's a wilderness kind of guy.

My nephew who is in the Marines thinks we seen a Buffalo or Bison - but there were no horns and this was not a four legged creature.

TIME AND CONDITIONS: Close to 11:00 to 12:00 dark with a misty rain. No street lights only our headlights. "It" was walking the same direction so we came upon it from behind. We used our peripheral vision to look - we were too scared to really take a gander.

ENVIRONMENT: Forest area hilly not very populated. There is a bridge nearby.

Follow-up investigation report by BFRO Investigator Eric Lester:

I spoke to the witnesses by phone, and found them both to be very sincere. The night they had their sighting was a rainy, foggy night, and when they first noticed it they were about a block away. They proceeded to slow down a little and as they drove by, they saw it peripherally, not wanting to look directly at it. The sighting lasted about 10 seconds.

It was described as definitely being taller than a normal human, with dark hair over its whole body. From what they could see of its head, it was 'sloping downward', with a huge jaw and a chin that almost touched its chest.

It moved very fast, with long strides to its gait. The creature never turned to look at them, and continued walking the whole time without moving into the forest.

They later had to stop and 'catch their breath', trying to figure out what they had just seen.

This sighting took place near Shades and Turkey Run State parks.

About BFRO Investigator Eric Lester:

++++++++++++++++++++++++
++++++++++++++++++++++++

YEAR: 2000

SEASON: Fall

MONTH: October

STATE: Indiana

COUNTY: Parke County

LOCATION DETAILS: Directions omitted upon request.

NEAREST TOWN: Marshall

NEAREST ROAD: highway 41

OBSERVED: My significant other and I lived in a house near (about 2 miles) from Turkey Run State park. It first started out, we were sitting near a fire we had built in our yard when I kept getting the feeling I was being watched..This was probably 9 or 9:30 at night. My partner asked me what was wrong to which I told him. A few minutes later, The cattle pastured behind our house started bawling and running like something either scared them or was chasing them. My partner stepped out to the side of the garage (Which was beside the house) and called me to come look, which when I got there, there was something tall standing in the moonlight. He then picked up a piece of steel pipe, his weapon of choice, and threw it at the shape, which did'nt move and we then assumed it was a pine or cedar tree. We went in the house shortly there after, but I still could not shake the " being watched " thing.

The next morning, we went back out to where the tree was supposed to be and there was NO tree !! He threw a big rock at the spot where he'd threw the pipe the night before and actually hit the pipe on the ground with the rock. We walked out to where these items were and the grass was all beat down and trails led out through the pasture where the cattle are. There was two distinct trails, the grass was waist high and you could tell the one whatever used to come up and the one it used to go back by the direction the grass was laying.

ALSO NOTICED: I remember it like it was yesterday. He does too.

OTHER WITNESSES: my partner. He was putting wood on the fire and asked me why I kept turning around and looking over towards the garage

OTHER STORIES: no..we moved to Terre Haute. Then later to Arkansas.

TIME AND CONDITIONS: It was cool and clear. There was a partial moon because we could see the the thing standing but not clearly.

ENVIRONMENT: forrests and open places. 5 strand barb wire fence 50 or so yards behind house.

Follow-up investigation report by BFRO Investigator Harold Benny:

I spoke with the witness by phone and found her to be credible. They had only lived in the house about three days when the incident occurred, and were not yet familiar with the surroundings. The eight foot tall tree-like object was standing immovable only fifty to sixty feet away in the dim moonlight, but was gone the next morning. Although there were two trails leading to the woods, no footprints were located.

Very close by is Turkey Run State Park. The origin of the name "Turkey Run" is unknown but the most accepted theory is that wild turkeys would congregate in the gorges (or "runs") for warmth where early settlers in the area would trap them in dead end gorges and hunt them with ease.

Photo taken at Turkey Run State Park -

About BFRO Investigator Harold Benny:

Harold Benny holds two degrees in Zoology and was a Biology teacher for several years. He participated in the following BFRO expeditions: Michigan UP 07, Arkansas I 07, Arkansas II (Oklahoma) 07, N. Florida 08, Missouri 08, Tennessee 09, and Ohio 09.

++

YEAR: 1998

SEASON: Fall

MONTH: September

DATE: 12

STATE: Indiana

COUNTY: Owen County

LOCATION DETAILS: Within Boundries of Wolf Cave Nature Preserve at McCormick's Creek State Park near Spencer Indiana.

NEAREST TOWN: Spencer

OBSERVED: A very low moan/howl was heard with a duration of 5 or 6 seconds. This initial moan was followed by a dog barking and a number of shorter low grunt like sounds. The sound was at an estimated distance of 100 yards to 500 yards northwest of our position.

OTHER WITNESSES: Myself 28, a brother-in-law 33, and our two 4 year olds (boy and girl) decided to take a nighttime hike to Wolf Cave which is about 3/4 mile from the family camping area of McCormicks Creek State Park. Our group hiked to the cave and briefly explored the mouth of the cave. The children wanted to go into the cave, but the dad's not being in the mood to crawl around in the dirt at night, decided no one would go far into the cave. We sat in the cave mouth talking for about ten minutes and then decided to hike back. About a tenth of a mile from the cave the moan was heard by the two fathers in the direction of behind us to the right. I initially heard the sound quite clearly but not wanting to alarm the children I said nothing. About two seconds after the noise my brother in law, a LT. in the US Army, asked me what the sound was. I replied "I don't know". We then heard a dog bark farther away and some short deep noises in the vicinity of the first moan. He then asked me if I thought the sound was made by an animail and I again replied "I don't know". About 10 steps later I suggested we put children on our shoulders and "make some time". He agreed. No further sounds were heard.

TIME AND CONDITIONS: The time was approximately 10:30 p.m. CDT

ENVIRONMENT: The terrain inside Wolf Cave Nature Preserve is thick dedideous forest (mostly ash and oak).

Like much of southwest Indiana, the area is very hilly and filled with a many ravines, limstone rock out croppings and cliffs. The cave and surrounding area is quite heavily visited during the day. The cave itself is small and had nothing to do with the sound. The Owen State Forest borders the Park and the whole region is fairly roadless and borders the White River. When the sound was heard by our party we were walking up the side of a ravine near the back of the preserve

++++++++++++++++++++++++
++++++++++++++++++++++++

Thursday, May 01, 1980

Unusual Encounter with Large Hairy Animal

By Ron Schaffner

Creature Cronicles

Witnesses: Tom and Connie Courter
Henschen Road and Indiana S.R. 56
Ohio County, Indiana between Aurora and Rising Sun

Facts of Incident(s):

The Courter's had left their mother's house in their car and were headed back up Henschen Road to their trailer. Once home, Tom got out of the car first so that he could get the diaper bag out of the back seat. (The Courter's had a six-week-old baby, which was also in the back seat.)

As he turned back around, he heard a strong noise which sounded like an "UGH". As he looked up, he saw a large hairy animal about "18 inches" away.

In a later interview, he told me that the creature as being about 12 foot tall, black and hairy, with large red eyes. He further stated that the head was shaped like a human and that its arms were hanging to the ground.

Tom quickly jumped into his car and spun his tires, as the creature took a swing at him. Tom said that the creature hit his car. Both Tom and Connie were obviously very scared and they went directly back to their mother's house.

On the next night, they stayed at their mother's house until 11:45PM. This time, Tom had his 16 shot .22. They parked in front of their trailer for a while, when they saw the same animal standing next to a tree on the other side of the road.

Tom fired one shot at the creature, but missed. He fired several more shots. He said that the animal seemed to dive to the ground and then seemed to vanish.

The Courter's filed a police report, but the Ohio County Sheriff's dept were very skeptical. In fact, they told us that they believe the couple were on drugs.

The authorities stated that they could find no evidence to back up their claim.

Our group did find smashed vegetation a few days later at the sight. The side of their car was smashed and we found no evidence of rust, or other metal shavings or paint.

Both Tom and Connie were still very fightened as we interviewed them a few days after their alleged encounter. At the time, we could find no hidden motive as to any possible hoax

Please remember that the size and distances mentioned by the witnesses are probably exaggerated.

BIG FOOT IN INDIANA

OHIO COUNTY COUPLE, POLICE, DIFFER OVER EXISTENCE OF CREATURE

BY BILL ROBINSON

RISING SUN - Picture a hairy, ape-like creature 12 feet tall that makes "a real funny noise like an ugh" and you have Mrs. Connie Courter's eyewitness description of Ohio County's own "Big Foot."

"If my husband stood on my shoulders he'd still have to look up at it. And it wasn't a bear," Mrs. Courter, of RFD 1, Aurora, said emphatically.

One night, while her husband, father-in-law and two police officers [??] were shooting at the strange creature, she added, "We heard it holler all the way up there" (at the home of her parents a barely safe distance away.)

Two nights last week, Mrs. Courter, 20, and her husband, Tom, 18, said they were confronted by the strange creature as they attempted to leave their car and enter their mobile home on Henschen Road just west of Indiana 56.

Both nights, they had returned from visiting Mrs. Courter's parents down the hill and around the turn on Ind. 56.

In each case, the Ohio County Sheriff, Francis "Swede" Colen, dispatched his brother and only deputy, Ora "Oop" Colen, to investigate. Both times, Oop returned saying he found no trace of any "Big Foot."

Here's Mrs. Courter's story of what happened:

"Last Tuesday night, my husband, with our little baby (six weeks old), had been at mother's. We took my sister-in-law, Debbie Tinsley, home on Henschen Road, turned around in Debbie's driveway and came back down the hill and pulled in front of our trailer."

"I had my door open. He had his door open and he reached down in the back seat to get the diaper bag when - it was only 18 inches away - he jumped back in the car and yelled. 'Close the door!'"

Mrs. Courter said Big Foot crashed against the car and dented it. But they sped away and went to the home of her mother, Mrs. Betty Tinsley, and called Sheriff Colen.

The incident occurred about 11 p.m. The next night, the Courters also visited Mrs. Tinsley's and this time waited until 11:45 p.m. to return home.

Tom Courter came prepared. He was armed with a 16-shot .22 caliber rifle.

When they parked, Mrs. Courter said she told her husband, "I'm not getting out of the car." he open the door and then said, "Did you hear that!"

"It was a real funny noise - like an ugh - and then we saw him sitting perched on the hill," she said.

"He fired one shot at it and it jumped up. It started acting like it was going to leave and he fired all 15 shots left into it. It would crawl to get away from the shots," she said.

Bibliographical Information:

Creature Chronicles #1; Spring 1980
rschaffn@tso.cin.ix.net

++++++++++++++++++++++++++++
++++++++++++++++++++++++++++

YEAR: 1982

SEASON: Fall

MONTH: September

STATE: Indiana

COUNTY: Morgan County

NEAREST TOWN: MORGANTOWN

NEAREST ROAD: 135

OBSERVED: I WAS A SENIOR IN HIGH SCHOOL IN THE FALL OF 1982. THERE WAS A KID IN MY ACCOUNTING CLASS THAT WAS A JUNIOR. I DON'T KNOW HOW WE GOT ON THE SUBJECT, BUT HE STARTED TALKING ABOUT A BLACK BEAR THAT WAS RAIDING TRASH CANS ON HIS ROAD. THE INTERESTING PART ABOUT THE STORY WAS THAT HE TOLD ME THAT THIS BEAR WALKED ON TWO FEET LIKE A HUMAN

AND HE DIDN'T KNOW IF IT WAS A BEAR BUT MAYBE A BIG MONKEY. AT FIRST I TOOK IT WITH A GRAIN OF

SALT BUT EVERY WEEK HE SEEMED TO HAVE SOME STORY ABOUT THIS BEAR/MONKEY. I GUESS THAT HE GOT THE FEELING THAT I WASN'T TAKING HIM SERIOUSLY SO HE INVITED ME TO COME TO HIS HOUSE "ANYTIME" AND HE COULD CALL IT IN WITH A COYOTE CALL. I SAID HOW ABOUT TONITE! I KNOW FOR A FACT THAT IT WAS A FRIDAY LATE SEPTEMBER ABOUT 8:30 PM. I HAD ASKED ANOTHER FRIEND OF MINE TO GO ALONG WITH ME. I TOOK A RECURVE BOW AND A BOWIE KNIFE AND HE TOOK A 22.CAL RIFLE. WE WENT TO MY CLASSMATES HOUSE THAT WAS AJACENT TO MORGAN MONROE STATE FOREST. WE WALKED ABOUT A QUARTER MILE TO A FENCE THAT MARKED THE STATE PARK ABOUT 40 YARDS INTO THE WOODS(THE STATE PARK) MY CLASS MATE HAD BUILT A HUGE TREE PLATFORM THAT WAS ABOUT 8FTX8FT AND ABOUT 12FT. TALL. THE AREA AROUND THE STAND WAS COVERED WITH SAPLING

TREES SO THICK THAT YOU COULDN'T SWING A BASEBALL BAT. WE CLIMBED UP INTO THE STAND AND MY CLASSMATE BEGAN BLOWING THE COYOTE CALL (THAT SIMULATES A WOUNDED RABBIT).THE NITE WAS WARM WITH A LITTLE BREEZE AND QUITE PEACEFUL. THIS CONTINUED FOR ABOUT THIRTY TO FORTY MINUTES WITHOUT A RESPONSE. WE WERE ABOUT TO CALL IT QUITS WHEN WE HEARD SOMETHING WALKING IN THE UNDERBRUSH. I THOUGHT IT WAS A RACOON/DEER BUT IT STARTED TO CIRCLE THE PLATFORM AND THE FOOTSTEPS WERE MUCH LOUDER THEN A RACOON, AND A DEER WOULD HAVE SMELLED US AND RUN AWAY. IT SEEMED TO BE LOOKING FOR SOMETHING (THE RABBIT) AND WE HAD REMAINED SILENT. THEN I NOTICED EVERYTHING AROUND US WAS SILENT EXCEPT THE BREEZE . THEN IT STARTED GRUNTING! THE GRUNT WAS VERY UNIQUE AND VERY

POWERFUL VERY DEEP CHESTED! A
SOUND I HAD NEVER HEARD BEFORE,
AND NEVER HAVE AGAIN. WE TRIED
TO SHINE A FLASHLIGHT ON IT BUT IT
WAS ALWAYS RIGHT AT THE EDGE OF
THE LIGHT WE NEVER SAW WHAT WAS
CIRCLING US. THE CIRCLING FINALLY
STOPPED AND WE ALL HAD TO GET
HOME BUT THE AREA AROUND US WAS
STILL QUIET, ONLY THE BREEZE AND
THE RUFFLING OF THE LEAVES IN THE
TREES. THE THREE OF US WERE
SCARED TO DEATH ABOUT GETTING
DOWN FROM THE PLATEFORM! WE HAD
CALLED IT IN AND IT WAS EXPECTING
SOMETHING! WE FLIPPED A COIN TO
SEE WHO WENT DOWN FIRST. I LOST
AND MY BOW WAS WORTHLESS IN THE
SAPPLING COVER SO ALL I HAD WAS
MY KNIFE. DROPPING DOWN INTO
BLACKNESS(AND IT WAS BLACK) NOT
KNOWING WHAT WAS GOING TO
HAPPEN WAS THE SCARIEST THING I
HAVE EVER DONE IN MY LIFE! BOTH OF
THEM CAME DOWN RIGHT AFTER ME

AND WE ALL RAN FROM THE WOODS AND ACROSS THE FIELD TO MY CLASSMATES HOUSE. MY FRIEND AND I JUMPED IN HIS TRUCK AND THANKED MY CLASSMATE AND WENT HOME AND NEVER TALKED ABOUT AGAIN.

I HAVEN'T REPORTED THIS INCIDENT TO ANYONE EXCEPT A FEW CHOICE FRIENDS AND FAMILY. I KNOW THIS ISN'T A CLASS "A" REPORT BUT I THOUGHT THE USE OF COYOTE CALLS MIGHT BE OF SOME HELP. I TRULY BELIEVE "AND ALWAYS WILL" THAT WHAT CIRCLED US AND GRUNTED SEVERAL TIMES WASN"T A BEAR, COW (MY FAMILY RAISED 45 HEAD OF CATTLE YEARLY), OR A PRANK . WE WERE ARMED AND COULD HAVE STARTED SHOOTING ANYTIME IF WE WANTED TOO.

ALSO NOTICED: SILENCE

OTHER WITNESSES: THREE

OTHER STORIES: I HAVE HEARD, FROM AN OLD CLASSMATE OF MINE THAT USE TO TELL EVERYBODY THAT HIS DAD WAS ALWAYS FIGHTING A MONKEY THAT WAS STEALING HIS CHICKENS

BUT IT WAS A BIG MONKEY. I DON"T KNOW IF THIS IS TRUE BUT I HAVE HEARD IT USED IN SEVERAL CONVERSATIONS OVER THE YEARS. THEY ALSO LIVE IN THE SAME COUNTY AS MY REPORT AND NOT FAR AWAY FROM THE REPORT. I ALSO KNOW THAT A PROFESSOR AT PURDUE UNIVERSITY DID A RADIO SHOW ON BIGFOOT IN ABOUT 1996 ABOUT BIGFOOT ACTIVITY IN SOUTHERN INDIANA I DON'T KNOW HIS NAME BUT THE STATION WAS WFBQ-Q95 HOSTED BY JAY BAKER. THESE REPORTS INCLUDED AREAS VERY CLOSE TO MY REPORT. I ALSO HEARD REPORTS ON

THAT SHOW FROM SWEATWATER LAKE IN BROWN COUNTY.

TIME AND CONDITIONS: NIGHT FOREST COVER VERY DARK OUTSIDE FOREST LITTLE CLOUD COVER SOME MOON LIGHT

ENVIRONMENT: FARM LAND AJACENT TO FOREST I THINK IN THE FOREST WE WERE ON THE TOP OF A HILL WITH A CREEK BELOW

NOT SURE!

++++++++++++++++++++++++
++++++++++++++++++++++++

YEAR: 2009

SEASON: Summer

MONTH: July

STATE: Indiana

COUNTY: Montgomery County

LOCATION DETAILS: Interstate 74 runs from Indianapolis head west to the Crawfordsville exit and head south. Take highway 47 to 234 and head west about 5 miles and you will see the Shades sign.

NEAREST TOWN: Waveland, Indiana

NEAREST ROAD: State Highway 234

OBSERVED: The first incident was in Febuary of 2009 after a snow. Me and my daughter who was 5 at the time went to Shades State Park to play. I was goofing around and making tree knocks every 15 minutes or so. I did not expect anything but any time I'm in the woods I try. We were making our way back to the truck and were within 500 yards when I made another set of knocks wich was followed almost immediatly by a very loud and powerfull growl. My daughter immediatly looked at me and said what is that dad. Halving my 5 year old with me made me just want to get her to the truck asap. The power and intensity was really something, it reminded me of passing a snow plow on the road, just the power of it.

The following summer a couple of things happened I think are related. In june me and my buddy who like to whitewater kayak went to the Shades, Pine Hills area to kayak due to some heavy rains and flooding, wich in Indiana is the only good runs we get. That day while standing along suger creek at our buddys cottage which is across from the spot Indian creek enters suger creek we were debating how safe Indian creek would be to run when for some reason a deer jumped into the water and tried to swim across, it did not make it and got sucked under a log jam under the covered bridge. That was our sign and we descided tomorrow would be a better idea. We met around 10:30 the next day and Kayaked Indian creek which was still running fast but safer. As we was leaving Indian creek and entering suger creek at the spot we were at the day before except on the other side, I noticed to my left along the cliff face was a stack of rocks about a foot and a half tall stacked right on top of each other. What was odd was this area was still quite flooded and the area was completely under water the day before.

There is know person in there right mind that would halve been in that spot over night or that morning.

My last incident was in July 2009. I was catfishing about 1 am at lake Waveland about 1 to 2 miles from the Shades and Pine Hills area. I heard very cleary three short howls coming from the area around Shades. And the whole time this thing was howling every coyote in the area was going nuts. As soon as it stopped everything went silent, it was really cool I pretty much knew at that point for once and for all that sasquatch is a real and living creature, very cool. I got online when got home and found a identicle sound, the Columbiana County Ohio howls are exactly what I heard.

ALSO NOTICED: The rock stack was odd because there is know acsess to the spot it was and it was surounded by a cliff on one side and pretty nasty flood waters around the rest it was just a small point that had poked up out of the water as it started coming down. No way a person made it no way.

OTHER WITNESSES: The growl was me and my daughter she was making snow angels and I was knocking on a tree.

Catfishing was me and my buddy on his boat.

The rock stack was while me and a buddy kayaking.

OTHER STORIES: Yes this past week my mom called . Her and dad had taken the yorkie out to Shades for a walk. They noticed right away that there were not the usual bunch of birds and critters about. Also while they were walking atleast two animals were whooping back and forthe. We listened to the Sierra whoops and that was very close to what they heard.

TIME AND CONDITIONS: Summer and winter. The rocks were after a pretty nasty flood in June. The howls were 1 am in July. The growl was mid day after a descent snow in Febuary.

ENVIRONMENT: The most notable things are this area consists of Shades, Pine hills, Lake Waveland, and 10 miles south Turkey Run State Parks. Sugercreek wich is a large creek runs through the area to the Wabash River wich runs down to the Ohio. The area is very rural and undeveloped.

Follow-up investigation report by BFRO Investigator Eric Lester:

After speaking with the witness, the following details can be added:

-the growl heard by the witness and his daughter was possibly within 100 yards, and lasted about 6 seconds; they were hiking on a trail in Shades State Park that runs to Sugar Creek, then loops back to the parking lot

-the stack of rocks was made up of 4 rounded river rocks; the witness was emphatic that nobody could have been to that spot during the previous day to do anything like that, as the water had receded sometime during the night to expose this area...if the rocks were stacked previously, it is probable they would have collapsed due to the strong current

-there were a total of 3 howls, each lasting 3-4 seconds each, with a few seconds between howls; coyotes would join in halfway through each howl, then all would stop at nearly the same time...they originated from the north, towards Shades State Park

Shades State Park is bordered by Turkey Run State Park to the southwest, and Pine Hills Nature Preserve to the east. Sugar Creek runs through both state parks, eventually flowing into the Wabash river to the west. Though this is the first report from Shades state park, two reports from Turkey Run State Park are included in the BFRO's database:

+++++++++++++++++++++++++++
+++++++++++++++++++++++++++

YEAR: 1979

SEASON: Fall

MONTH: November

STATE: Indiana

COUNTY: Monroe County

NEAREST TOWN: Harrodsburg

NEAREST ROAD: Popcorn Road

OBSERVED: After caving in a local pit off of Popcorn Road in Monroe County I decided to wait on the surface while my two fellow cavers did the next pit.

I was sitting in the back of my jeep reading. I saw a silhouette moving south just within the tree line. Thinking this was the land owner I got of my jeep and walked up to the tree line.

The jeep was parked in a small grassy clearing used for camping and the two pits were appox 50 feet east of the jeep with the fence and tree line 10 to 20 feet beyond the pits. The trees were bare of cover and it was a bit overcast.

I waited at the fence line and I kept an eye on the silhouette moving towards me. I called out once, twice with no reply. The thickness and number of trees and the overcast made it hard to see details of the silhouette.

The silhouette moved past me and the only noise I heard was the slight crunching of leaves on the ground. Still not able to make out more then a silhouette and thinking this was the land owner or local, I called out again and waved at the silhouette. It was now approx 50 to 60 from me. A bit confused and thinking that the silhouette did not hear me nor saw me, I moved down the fence line to a small opening cut in the tress for a power line. I stood at the fence waiting, and the silhouette moved into the opening. I called out again and froze. The silhouette also froze and turned its head towards me. We stared at each other for 10 seconds or so. No other interaction took place. I then was able to view the silhouette and make out some detail.

The silhouette stood appox 6-7 foot tall, was covered in long dark brown fur/hair, and stood on two legs with a bit of a bend forward. After the brief stare the witness turned its head back and continued to walk back into the trees and I returned to my jeep.

Officially I can not say for sure it was a 'Bigfoot', only that it was walking up right on two legs, had two arms, a chest, and torso and was covered in long dark brown fur/hair.

ALSO NOTICED: Birds calling out, no other sounds or smells

OTHER WITNESSES: none

OTHER STORIES: Later years southeast of lake Monroe. South of my sighting.

TIME AND CONDITIONS: Mid morning, early afternoon

ENVIRONMENT: Wooden area with small clearing. Hilly. Farm land.

Follow-up investigation report by BFRO Investigator Caroline Curtis:

The witness and I have discussed this incident on the phone a few times and I found him to be very credible.

Thinking he was seeing the property owner or a person that didn't see or hear him, he moved to a position where he would see the person eventually come out of the woods. When it came out of the trees into the clearing he was able to see that it was not a person. He was close enough to know they were making eye contact but because of the overcast day and shadows, he was not able to see additional detail of the face. He did not feel threatened and he did not get the sense that the creature was too concerned about him.

His friends were still in the pit. Still working for an organization that maps caves and their entrances, they have documented a total of 3,434 caves in the state of Indiana. The difference between a pit and a cave is a pit is vertical and a cave is horizontal.

The location was next to Clear Creek and only about five miles from Lake Monroe (actually an artificial reservoir). The lake is the largest in Indiana with 10,750 acres, including three recreational areas - Fairfax, Hardin Ridge and Paynetown. Indiana's only federally protected U.S. Wilderness Area, the 13,000 acre Charles C. Deam Wilderness Area, is located on the

YEAR: 1982

SEASON: Spring

MONTH: May

STATE: Indiana

COUNTY: Monroe County

LOCATION DETAILS: stinesville go north on gravel road to 3/4s of mile where old road bridge used to be. Turn right thru dirt in field go as far as you can then walk another mile or so.

NEAREST TOWN: Gosport Ind

NEAREST ROAD: Moon road

OBSERVED: Back in 1982 a friend, his brother and I was fishing along the White River about 4 miles north of Stinesville. We had been there many times before. That day I happend to mention that there was no birds singing or had noticed any game there. After a few minutes my buddy said that he hasnt seen any either. He did say he had a feeling of being watched ,and his brother had said the same thing. After awhile of looking around while we fished I looked up on this tall ridge behind us about 50 to 75 yards behind us I thought I saw "someone" watching us. I pointed it to my buddys and they saw it to. I have lived in the country all my life, hunted day and night, still I never felt this way before. We watched each other for a long time, it moving so carefully trying to get a better look .After 45 minutes or so we thought it best we leave so we started back to my truck and "it" followed us from a distance never getting out in the open but not really trying to hide that well either. we also noticed a faint smell of something dead or so we thought. We made it to my truck and didnt take too much time loading it, after all we walked about a mile and was ready to leave. We still tell this story today and not sure what it was but "Bigfoot" has been mentioned not disputed by my buddys or I.

ALSO NOTICED: I have been back many times later and havent noticed any thing like that again

OTHER WITNESSES: 3 we were fishing

OTHER STORIES: When I was in school the kids that lived in the area once in a while would tell of something strange, like strange noises, thought they saw monsters things like that but not often. But I heard more stories from the lake monroe area than where we were.

TIME AND CONDITIONS: around midafternoon bright clear day

ENVIRONMENT: hilley, no houses to bother you ,along the river side ,very wooded and lonely place ,not well travled on foot or ride

south shore.

++++++++++++++++++++++++
++++++++++++++++++++++++

EAR: 1988

SEASON: Winter

MONTH: November

STATE: Indiana

COUNTY: Monroe County

LOCATION DETAILS: Hindustan is the nearest village (only a few homes and small church) perhaps 3-4 miles away on Old 37. The town of Martinsville is about 5 miles to the north and the town of Bloomington is about 15 miles to the south. This incident took place in the Morgan-Monroe State Forest on the Main Forest Road about 1 or 1.5 miles east of the forest entrance that is off of the State Road Old 37.

NEAREST TOWN: Hindustan

NEAREST ROAD: Main Forest Road

OBSERVED: This took place while on the last day of firearm deer season in Novemeber, 1988. It was after sunset but not yet dark. The temperature was in the mid 30's but the ground wasn't frozen and there were still patches of snow on the ground in places. I came out of the woods, put my shotgun, heavy winter coat, and daypack in the trunk of my car and got inside my car to start it up when a movement to my left caught my eye. I will tell you what I saw and what I thought as I was seeing this happen. This man/bigfoot came out of thick brush between the trees to my left on the opposite side of the road from where I was parked, maybe thirty five or forty yards down the road at an angle but it had to know that I was there or at least that my car was there. When I first saw this thing coming out of the thick brush the first thing I thought was, where in the heck did he come from.... Then, I thought, stupid guy, no hunter orange and then I began to notice strange things like he had no gun, and that he was the same color from head to toe, no hat, no hunter orange? But then what I really noticed was that his arms were long, his hands were swinging at his knees and though he was taking large steps crossing the road very quickly but without running, he was just walking. His arms were swinging wide like a British soldier marching but he didn't seem to be in any hurry and he completely ignored me like I wasn't there but he had to know that I was there. He did not look over and kept his head looking straight ahead. He went into the woods on my side of the road (side I was parked on) and went down the hill and disappeared. My eyes told me that I just saw a bigfoot but my brain and common sense was trying to dispute that. His color was dark brown and I thought could that have been a man in a one piece Carhartt outfit or if he were a hiker lost wouldn't he just have stayed on the road rather than go down a steep hill when it was almost dark or if lost why didn't he ask me for help? He was about six feet tall but very husky with his hands, body, feet, side of his head-face, all one color. I actually couldn't see fur because the light was dim. I could see the trees and tree trunks fine but not the bark because the tree trunks were silhoutetted and that's what I mean about not seeing the fur, it was all one shade of dark brown. Now, all of this that I just told you took place in a minute or less. Soon as it disappeared over the side of the hill I jumped out of my car, opened the trunk and loaded my shotgun (for self-defense only), put on my winter coat, and got my flashlight out of my day pack and walked to where I thought the thing went over the hill. I wanted to check out what I just saw and was hoping to find a footprint in the ground or on a patch of snow and I went slowly back and forth, traversing this rather steep hill going a little deeper down as I moved back and forth. There was a deep like gorge on the right side of the hill and I looked up towards the road and by now the ridge on top was just a slither of gray and looking over my shoulder I realized this was not too smart, it was pitch black below me and I got a little scared and hiked back up to the road and to my car as fast as I could. I placed my shotgun back in the case along with my heavy coat putting them in the truck and then I drove up and down the main forest road to see if there were any cars parked off on the sides or if there were anyone walking, I saw nothing. I drove to the valley road where I thought this thing would eventually come out at the bottom of the hill into farm fields and I drove back to the end of this valley and back hoping to see something but saw nothing. I have been wanting to make a report of what I saw and have told my wife, children, teenage granddaughters, brother, and two close hunting buddies, but have always been hesitant to make a report. The truth is, though I am sure what I saw wasn't a man I can't prove it wasn't a man dressed up in a suit but who would be that stupid especially during hunting season. I can't say positively that it was a bigfoot but that is what my eyes saw. Logic says "no", my eyes say, "yes" and that's my report the best that I can remember it. I am glad to get this off my chest. If this helps in any way then that's great but please use only my first name if this is posted,

I am not seeking notoriety here, thanks. Sincerely, [name removed]

ALSO NOTICED: I came back the very next day to see if I could find footprints in the soft ground (from snow melt) or on remaining patches of snow but saw absolutely nothing and as large as this creature was I was hoping to see, in the daylight, at least something but saw nothing?

OTHER WITNESSES: I was alone-no other witnesses.

OTHER STORIES: Only the two sisters report, BFRO #354 that I read on this site only their location is about five miles away from my sighting near Lake Lemon and their discription of size and color of fur did not match what I saw. However, I have heard about reports of sightings and was told that a team of students and professors from IU actively checked several reports of sightings on Hickory Ridge in Hoosier National Forest south of Monroe Reservoir which is probably 25 miles south of my sighting. These reports were back in the early 1980's if I remember correctly and there were a rash of sightings during that time including at Williams Dam area which I checked out myself but that's another story.

TIME AND CONDITIONS: Evening at dusk and the light was dim but not dark but trees were more silhouetted which made the bark hard to see, all one color in other words.

ENVIRONMENT: Heavy brush, thick woods on one side of road, flat blacktop forest road, trees (tall timber) steep slope or hill with gorge-like gully running down on the right on the other side of this road.

Follow-up investigation report by BFRO Investigator Eric Lester:

I spoke to the witness over an hour about his experience. He was initially reluctant to submit a report, but did so after family members urged him to do it. He recalled his experience vividly, and was able to recount details like it happened yesterday.

The creature came out of a thick-undergrowth area of the forest, and crossed the two-lane road in about 3-4 steps, 'much faster than [the witness] could'. Physical attributes of the creature that really stuck in his mind were the long arms/hands, the uniform brown color, little or no neck and the thick 'wrestler-like' build.

When he saw this creature, he was in denial that it could be a sasquatch, and has tried over the years to justify it being a human. He hasn't come to that conclusion, but has recently accepted that what he saw that day was probably a sasquatch.

About BFRO Investigator Eric Lester:

++

YEAR: 1989 or 1990

SEASON: Fall

MONTH: October

STATE: Indiana

COUNTY: Monroe County

NEAREST TOWN: Unionville

NEAREST ROAD: Lakeshore Drive ?

OBSERVED: In southern Indiana, Unionville October 1990 or 89. My sister and I were on our way home late at night (approx. 2:00 a.m.) when a bigfoot walked in front of my sister's Honda touching the front of the vehicle. It was very hairy and had long arms. It walked upright, but this creature had a sideways gait as though limping or injured. It looked straight at us and walked slowly. It did not appear to be afraid. It was approximately 5 foot four I'd say. It closely resembled an orangutan.

TIME AND CONDITIONS: It was dark, but the creature was within feet of us. It was cool out, but I don't recall rain.

Follow-up investigation report by BFRO Investigator Jim Osborne:

I spoke on the phone with the primary witness and then later to her sister. Although in the report the witness stated that it looked like an orangutan, when I directed her to a website with pictures of orangs she stated that it didn't really look like that. Her sister stated that she was in such a state of shock when it occurred that she really didn't get a good look at the face. Her estimate of its height was 4 1/2 to 5 feet tall. The witness said the hair was orangish-brown, while her sister called it orangish-red.

(This area is very near the central Indiana Hoosier National Forest.)

++++++++++++++++++++++++++
++++++++++++++++++++++++++

YEAR: 2009

SEASON: Spring

MONTH: March

DATE: 16

STATE: Indiana

COUNTY: Monroe County

LOCATION DETAILS: Bloomington location removed per witness

NEAREST TOWN: Bloomington

NEAREST ROAD: Not sure

OBSERVED: I have had two possible "Bigfoot related" incidents. I've wanted to discuss the older incident for quite some time but have hesitated to do so. The second incident occurred March 16, 2009 and unnerved me enough to go ahead with these reports.

(Note: for the older incident, please see report #25793)

This occurred on the western outskirts of Bloomington, IN. While this happened in a subdivision, it is an isolated subdivision surrounded by fields, small growth forest and quarries. Only a few miles further west the forestation thickens and several creeks and streams run through this area. Deer are seen here frequently, as well as other game animals and occasionally turkeys.

My girlfriend and I arrived at her parents' home in this neighborhood. We exited the car and I immediately heard a very, very odd sound. They were short moans, about 2 seconds apart at first, then many different timing variations. But they seemed to come consistently in groups of 4, with each call lasting between 2-10 seconds. After my initial encounters with what I came to believe was a Bigfoot, I found the BFRO site and listened to all the sounds, viewed all the images, read as many reports as I could -- I became an avid fan of the investigation and the science behind the creature. I recognized this sound as similar to the Ohio moans but shorter in duration. I'll have to listen to them again to make certain, but these were not anything like the whoops, screams or "bionic bird calls" that I've heard. These were definitely loud, powerful, reverberating moans. I have heard owls call out responses to environmental noises and also heard coyotes howling many times -- you can tell the difference based on the sheer power of the tones. This was not an owl or coyote. My girlfriend likened it to a very sick or injured cow -- but none are near this location. She had heard sick cows before and this was not a sick cow mooing -- it was, again, definitely a moaning sound.

We listened for 2 or 3 minutes then went inside. She immediately went out the back door to listen, as she knew that I would want to do so. We listened another few minutes and I decided to try and call back. I first tried mimicking the moan and, not receiving a response, I tried to make a whoop sound as best I could. The moaning stopped immediately. Off to our right, about 200 yards away in a stand of trees, I could see a few trees visibly shake -- not wiggle in the wind but shake like an elephant had just moved between them. I heard no noises like breathing, growling, grumbling, footfalls or otherwise. However, at that moment, off to our right a dog started barking ferociously and we could see it backing up closer to its house. This continued down the row of houses -- the dog would bark, growl then back up -- and it was in succession, getting closer to us. I was very scared at this point as, even if it was not a Bigfoot, I'd apparently gotten its attention and it was moving towards us. After passing the barking dog 3 houses away on our right, the dogs silenced and everything fell silent. At that moment, I saw some movement out in the field to the west of the trees, about 50 feet beyond the stand of trees. I heard a couple of quick moans then something streaked across the field from north to south. To describe what I could see is difficult. The moon and star light was enough that I could see a shape and a grayish color. The figure was definitely on two legs. It was running very fast but the legs were taking very large strides while the body appeared to be kept almost parallel to the group, like you would run if someone were shooting at you. The figure did not appear to be much taller than a normal person though it did seem (again, in poor lighting conditions) to be stockier than an average man. The shoulders were much larger than the waist and the legs were long and appeared lean but strong. The trunk of the being was too blurry to see any detail. The creature took 4-5 long and power strides and dropped to the ground behind what appeared to be a bush or clump of weeds. At that point, I felt nervous enough that we went inside. I peeked out several more times in the ensuing few minutes and saw nothing. I had been kicking myself about not having a camera, but the lighting was so poor and the movement so fast that an indistinct gray blur is all that would have turned out. About 10 minutes after all of this ended, we heard the sound of metal clanking and being broken. My girlfriend advised that there was lawn furniture further out in the field and it sounded, to her, like someone was throwing or breaking it. I looked a little later and yes, in fact, the lawn furniture had been strew about when it had been sitting upright when we arrived home (I didn't see this as I am not all that familiar with the setting -- she noticed it was upright because she knew what/where to look at). There were no storms, wind or other natural disturbances in the area that night, though I cannot preclude a common animal, such as a deer, may have run through the area and knocked everything over. Other than being on edge the rest of the evening, waiting for something to happen, nothing else was noted.

In talking with her further about this, I learned that the neighborhood is aware of this creature. They have nicknamed it "The Troll" and generally give it a wide berth and respect it. They say it lives in a sinkhole further out west, but

I'm not certain exactly where. I've asked her family to take me out to the sinkhole so I can take a look but have not made the trip yet. Sightings of this creature are seasonal, though I don't know which seasons/time frames it is commonly present in.

This second sighting felt a little more threatening, or at least I had the vibe that the situation was not safe or controlled (from either side -- mine or the creature). I got the distinct impression that this was a juvenile calling out and when he got an unexpected response, he came to investigate, chest puffed out and ready to rock. But he quickly backed off and instead sought to either hide or make his way across the field and away from us. When we went inside, I think he showed his displeasure by trashing the lawn furniture. I could easily be completely wrong in my assessment, but those were the feelings I got from how things played out.

ALSO NOTICED: Nothing out of the ordinary either time

OTHER WITNESSES: 2 witnesses, only I am willing to discuss at this time, unfortunately.

OTHER STORIES: Bloomington -- many stories of Bigfoot-like creatures in the area.

TIME AND CONDITIONS: between 10PM and 2AM. Some starlight, low moonlight or cover blocking light. Weather was cool and clear.

ENVIRONMENT: A field near thick forest. Not familiar with the area beyond that.

Follow-up investigation report by BFRO Investigator Eric Lester:

I spoke to the witness at length about his experiences, and decided to separate his 'older' experiences into a new report (see report #25793).

The moans were fairly close to them, within 300 yards. He seemed to think the moans were from one individual.

As the creature moved through the forest, he noticed the trees shaking, and it was about 100 feet away from them at the closest point. The thick brush area that the creature ran into was about 70 feet away. The only thing he noticed after it moved into this spot was possibly a head pop up, then down very quickly.

The witness states that some of the neighborhood residents refer to there being a "troll' in the neighborhood. He couldn't add much detail to this description, for none was ever given to him by anyone. Its possible that other sightings have occurred in this area, with the witnesses choosing to not look into it further.

After speaking to the witness about his experiences, I believe him to be truthful and sincere.

++++++++++++++++++++++++++
++++++++++++++++++++++++++

YEAR: 1981

SEASON: Fall

MONTH: October

DATE: 26

STATE: Indiana

COUNTY: Lawrence County

LOCATION DETAILS: It was just off the airport road east of Bedford and just south of Hwy 50 about 3 to 4 miles on a small gravel road near Leatherwood creek.we had just crossed a bridge over the creek and were heading east.

NEAREST TOWN: east of Bedford In.

NEAREST ROAD: south of hwy.50

OBSERVED: This sighting occured in october of 1981.I had just been discharged from the marine corps.I was 21 years old at the time.I have lived in this area of indiana all my life.I am an avid hunter and outdoorsman with much experience .I am a raccoon hunter also and have been hunting at night by myself many times.It was a warm Saturday afternoon in late october.My sister and her boyfriend and his sister and myself went on a picnic at Spring Mill State Park.After the picnic my sisters boyfriend and her went home.The other girl and myself decided to go out riding around in the country to enjoy the fall leaves. We were heading east of Bedford, Indiana on a small gravel road near the airport .This road went to a small creek bottom by leatherwood creek.There was thick forest on one side of the road and a cornfield on the other,surronded by thick briars.We had crossed the bridge over leatherwood creek and i pulled my 4x4 truck off to the side of the road,we were headed east.I was looking down adjusting the radio when my companion stated that someone was watching us.I looked up and to my disbelief I saw something standing in the middle of the road about 50 to 60 yards to our front.This thing was at least 6 to 7 feet tall standing upright.It was cover in long dark gray and brown hair.I could not see its face.I could feel the hair on the back of my neck standing up.My companion was terrified.I slowly removed an M1 carbine from its case and inserted a magazine the creature continued to watch us closely .I opened the truck door and stepped out on to the road.I began to move slowly towards the creature as I did it turned and ran thru the briars in to the cornfield.I returned to the truck very shaken I have never seen anything like this my companion insisted that I take her home .Which I did we never went out again.I have often wondered why I did not fire my weapon at the creature. I guess it looked somewhat human like and it made no threating behavior.I never reported this incident because I figured no one would believe me anyway.

ALSO NOTICED: none that i can think of

OTHER WITNESSES: just myself and my date we were riding around in my 4x4 enjoying the fall leaves

OTHER STORIES: noted some reports on your web site about encounters near bedford in the hoosier national forest which is very near to the location where these events took place.

TIME AND CONDITIONS: Late afternoon It was warm and dry very sunny

ENVIRONMENT: a creek bottom with heavy forest on one side and a cornfield on the other side surrounded by thick briars

----

Follow-up investigation report:

The first thing that entered Derek's mind when he had his sighting was who is that? Upon getting closer and hearing his friend's question repeated, he told me, "It was more like what is that?" He approached close enough to see the wind blowing the creature's hair around. I asked about the general appearance of the animal and was told "I couldn't see the eyes because that area was very dark and covered with hair."

He mentioned that the head did suggest being pointed or almost. The animal was thickly built with heavy legs. He mentiioned it had no neck. The color was a brownish gray. When he approached it he mentioned two things to me. The first was he didn't think he could shoot. It was so manlike and it wasn't "doing anything."

Secondly, if he used the weapon he had along it would have had a minimal effect which he wouldn't have felt comfortable about. He was just out of the Marines and full of energy during the time he had his encounter. Upon the approach of it he mentioned it took two quick strides and it was gone. Although manlike, it seemed to have the longest hair on its trunk (body). Also, he thought it might be somebody playing a trick on them since it was so close to Halloween. What changed his mind quickly was the way it moved and where it went. It leaped into a series of wild malty... rosebushes with thorns. It tore right through and it wouldn't be something that a person would do in Derek's opinion especially with the length of stride and speed exhibited. He also mentioned smelling an odor. This area was on the edge of the Hoosier National Forest.

++++++++++++++++++++++++++
++++++++++++++++++++++++++

YEAR: 1987

SEASON: Fall

MONTH: October

STATE: Indiana

COUNTY: La Porte County

LOCATION DETAILS: This was located just outside of the Kingbury Fish and Game Wildlife refuge

NEAREST TOWN: Stillwell

NEAREST ROAD: highway 104

OBSERVED: In October of 1987 a friend and I went camping just outside of Kingsbury Ordinance Plant (KOP for short) in Kingsbury, IN. We were about 1 to 11/2 miles off Hupp rd (which is the road the enters KOP). We set up camp just before dark around 6 pm or so. After about 45 min, my friend went home to eat dinner and call another friend of ours to come out when he gets off duty. Just around dusk, while getting fire started, I heard rustling in woods about 25 yards away. I didn't think anything of it at the time and because of the population of Deer out in that area, I thought it might be one. After about 10 minutes, heard noise again, but it was about 15 yards away this time. I yelled HEY real load and the noise stopped for about 5 minutes. Then I heard grunting about same distance away. Getting nervous, I yelled again Knock it off. I then picked up a stick and said, I have a gun. The noise stopped after 3 or 4 sec and as I turned to look at fire, I heard a loud thump to my right. I looked and saw a large tree stump that landed about 8 ft away. I turned around and faced woods and yelled very loudly Knock it off or I'll shoot. I then saw another large log come sailing through the air toward my direction and land about 3 feet in front of me and about 5 feet to left. I started running down a trail toward the road. I did not hear anything for about the first 100 yards, and then I heard rocks moving as if someone was running in the same direction as I, but they were on the RR tracks to my left. The sounds of the steps were about 1 sec. apart. I heard this for about 8 to 10 sec. then it stopped. I ran for about 25 more yards and stopped. It was dark and overcast skies, but still there was enough light to see where I was running. I turned and looked behind me and then to my left and saw nothing. I then turned back toward the road and focused on the street light ahead. Walking very fast now, I came to a dip in the trail. This is where I saw it, a large figure standing in front of me. The dip in the trail sloped down to about 3 feet deep and ran the depth for about 30 ft. then it rose again to ground level. The figure was standing in the center of the dip in the trail and stood at the same height as myself. I am 5'9" so I the figure had to be from 8 to 9 feet tall. This thing let out the most God forsaking scream, that it would make your blood curdle. I did what any normal person would have done; I took off running toward my right through an open field. I ran as fast as I could away from this figure. I finally reached my friends front steps about 3/4 miles away. At first I thought it was him, but when he came to door he had been eating dinner, his shoes where off, and his mom said that he had not left the house. After he ate and our other friend arrived we all went back out there about 3 hours later. The small tent we put up had been tore down and logs and stump that were thrown were lying in the spots that they landed.

I have never forgotten this. For years I have dreams, (Nightmares I call them) about the creature and that terrible scream. Till this day I am still scared to go out there.

ALSO NOTICED: I have heard stories of things that have happened in the area around KOP, but nothing that I can remember.

OTHER WITNESSES: No physical witnesses to the incident itself, but my Brother, and my friend saw how scared I was after this happened. My brother said that I was pale and still crying the next day when he came home.

OTHER STORIES: Nothing

TIME AND CONDITIONS: It was around 7 -7:30 pm

It was dark and overcast skies, but still there was enough light to see where I was running.

ENVIRONMENT: There were woods near my location and fields behind the woods. KOP was to my right and was a densely wooded area.

Follow-up investigation report:

Several phone conversations with the the witness's brother, and the witness himself, provide a consistant and credible report. The witness is still bothered with dreams of the incident.

Midwest Investigator, YEAR: 2008

SEASON: Fall

MONTH: October

DATE: 8

STATE: Indiana

COUNTY: La Porte County

LOCATION DETAILS: Willing to take to location for verification of story. Property belongs to friend and is my own private hunting grounds, so I do not care to turn it into a carnival.

NEAREST TOWN: Mill Creek

NEAREST ROAD: County Road 800

OBSERVED: I was deer hunting in a small woods that bordered a swampy area. The entire woods is on a 45 degree angle leading down to the swamp except the small bowl like section I was hunting in. The spot is very thick with pawpaw trees which provide ample cover for wildlife till about mid November. I was sitting in my tree stand about twenty feet off the ground. It was after 5pm which is what I refer to as magic time because that is when the deer start to move in the evening. As the last rays of light had begun to fall onto the forest floor two does shot out from the brush behind me. They seemed nervous and kept looking back behind my tree. I thought maybe a buck was following them so I nocked an arrow and prepared to see a large buck walk out of the brush from behind me. Then it hit me, the ungodly stench. It kind of smelled like garbage or like someone had lifted the lid on a septic tank. It was awful. Then a log flew out from the brush behind me and struck the tree I was in at the base. I nearly soiled my britches. I turned to look behind me and what ever it was it ran off busting through the underbrush and I never got a clear look at what it was. It let out a bloodcurdling scream as it ran off. I froze not believing what had just happened. I waited for at least an hour before I got down to walk back to my truck.

OTHER WITNESSES: Only myself

TIME AND CONDITIONS: Between 6pm and 7pm as the last rays of sunlight touched the ground.

ENVIRONMENT: Hardwood forest bordering a swamp.

Follow-up investigation report by BFRO Investigator Eric Lester:

After speaking with the witness about his experience, the following details can be added:

- it was only a number of minutes between noticing the stench and the log hitting the tree; the does were very nervous during this time

- what he saw moving through the brush as it ran off was 'at least as big as a human', though he could discern no details

- the log that was thrown at his tree was rotten, about 2' long by 1' wide

- the scream was heard from the far end of the woods, about 45 seconds after the creature ran off; he says it would have taken him about 5 minutes to walk to the area where the scream sounded from

- he described the scream as a 'deep tone', very 'unique', and rising in pitch near the end of the scream

- the area of the forest he was in is described as a 'travel route' for wildlife, being small and narrow

- terrain is very wet, almost swampy in areas

The Pawpaw trees mentioned are small clustered trees with large leaves and fruit, native to North America. This type of tree includes the largest edible fruit indigenous to our continent. They are understory trees in hardwood forests found in well-drained deep fertile bottomland and hilly upland habitat.

K.C. Charnes

++++++++++++++++++++++++
++++++++++++++++++++++++

YEAR: 2008

SEASON: Fall

MONTH: October

DATE: 8

STATE: Indiana

COUNTY: La Porte County

LOCATION DETAILS: Willing to take to location for verification of story. Property belongs to friend and is my own private hunting grounds, so I do not care to turn it into a carnival.

NEAREST TOWN: Mill Creek

NEAREST ROAD: County Road 800

OBSERVED: I was deer hunting in a small woods that bordered a swampy area. The entire woods is on a 45 degree angle leading down to the swamp except the small bowl like section I was hunting in. The spot is very thick with pawpaw trees which provide ample cover for wildlife till about mid November. I was sitting in my tree stand about twenty feet off the ground. It was after 5pm which is what I refer to as magic time because that is when the deer start to move in the evening. As the last rays of light had begun to fall onto the forest floor two does shot out from the brush behind me. They seemed nervous and kept looking back behind my tree. I thought maybe a buck was following them so I nocked an arrow and prepared to see a large buck walk out of the brush from behind me. Then it hit me, the ungodly stench. It kind of smelled like garbage or like someone had lifted the lid on a septic tank. It was awful. Then a log flew out from the brush behind me and struck the tree I was in at the base. I nearly soiled my britches. I turned to look behind me and what ever it was it ran off busting through the underbrush and I never got a clear look at what it was. It let out a bloodcurdling scream as it ran off. I froze not believing what had just happened. I waited for at least an hour before I got down to walk back to my truck.

OTHER WITNESSES: Only myself

TIME AND CONDITIONS: Between 6pm and 7pm as the last rays of sunlight touched the ground.

ENVIRONMENT: Hardwood forest bordering a swamp.

Follow-up investigation report by BFRO Investigator Eric Lester:

After speaking with the witness about his experience, the following details can be added:

- it was only a number of minutes between noticing the stench and the log hitting the tree; the does were very nervous during this time

- what he saw moving through the brush as it ran off was 'at least as big as a human', though he could discern no details

- the log that was thrown at his tree was rotten, about 2' long by 1' wide

- the scream was heard from the far end of the woods, about 45 seconds after the creature ran off; he says it would have taken him about 5 minutes to walk to the area where the scream sounded from

- he described the scream as a 'deep tone', very 'unique', and rising in pitch near the end of the scream

- the area of the forest he was in is described as a 'travel route' for wildlife, being small and narrow

- terrain is very wet, almost swampy in areas

The Pawpaw trees mentioned are small clustered trees with large leaves and fruit, native to North America. This type of tree includes the largest edible fruit indigenous to our continent. They are understory trees in hardwood forests found in well-drained deep fertile bottomland and hilly upland habitat.

++++++++++++++++++++++++
++++++++++++++++++++++++

YEAR: 2003

SEASON: Fall

MONTH: October

DATE: 11

STATE: Indiana

COUNTY: Kosciusko County

LOCATION DETAILS: This is a private residence on a small lake about 1+ miles north of town. Just to our east by a mile is a 4000 acre state wildlife fish and game area.

NEAREST TOWN: North Webster, Indiana

NEAREST ROAD: Just west of State Road 13

OBSERVED: Ok... not "just a story"... this happened to us recently here "at home" one night... Friday night, Oct. 10th, 2003 - Saturday morning, aprox. 1:00am.......

My son and two of his friends were coming home late one night, just about 1:00am on a Saturday morning after recent home-coming activities. Since their plans had gotten changed, I wasn't expecting them home; they were supposed to stay somewhere else. When they pulled into the drive/yard, I was already in bed halfway sleeping, when I awoke to the sound of someone pounding on the front door and yelling excitedly. Then I heard my name and "dad" coming from them. So, I slowly pulled myself from bed and made my way to the front room where I heard the sound of one of the bedroom windows sliding open. My son came falling thru and ran to the front door just about the same time I reached it so we opened it and let his other two friends in. They came in very quickly. In a jumbled mess of excitement, they started telling me about "it" being "down there and coming up here" and all sorts of things I was too groggy to understand. Saturday after work, I finally sat down with two of them to hear it all again and get the whole story, in order, and without all the confusion there had been the night before. Here's how they explained it....

After pulling into the yard up near the trailer, they walked in the dark up to the porch and started to knock. Friday night, the 10th of October was right around a full moon, and the sky was absolutely clear that night. As they walked up on the porch to knock, one of the three looked back over his shoulder 'coz he heard something and noticed a "stump" down in the yard, about 150' away, near the boats at the edge of the water. The other two looked then to see what he was talking about, when "the stump" got up and just stood there. Even in the moonlight, they could see it was NOT a deer... not an "anything" except the form of a person, 'coz it stood there directly facing them and they could see it easily. So as they started to knock more excitedly, it began to walk away quickly to the east on the shore line, but then abruptly turned around and began to move much quicker right back across the yard where it had been and toward the marshy area and light woods and tall grass immediately south of the porch and trailer. When my son saw this he came running around the trailer and let himself in through the window. Now the other two could hear the sound of "this guy?" running heavily into the tall grass, sticks, branches and all the stuff that was down in the woods between the trailer and the lake, but then heard it starting to move closer in up the hill. That's when the hard pounding and yelling to let them in really started. We let the other two in, locked the door... I went back to bed, while they stayed up half the night rehashing what had taken place and how it freaked them out.

4 hours later, I was up... ready for work and out the door around 5:25am. I had my son get up and briefly talk me thru what they had seen earlier. Even at 5:30am as I was leaving, the moon had made it mostly west in the early morning sky and was still so bright I could easily make out every tree, patch of grass, boat at the shoreline... every little thing we're used to seeing out here. Back at 1:00am when this happened, it was even much more illuminated with the moon directly overhead, especially with the light bouncing off the water as well.

So, maybe it was "just some guy" out squatting and then standing in 5 inches of lake water at 1:AM in the middle of the country? Well, when I'm 6 foot and about 200 pounds, and they see some "guy" who's way taller than me and a lot larger... and built a LOT larger... way taller... way larger... and grunting slightly as he ran...?? We only have one neighbor, and knowing them, they don't run around in a yard of standing lake water at one in the morning... or Noon for that matter. He's also not approximately in the range of 8 feet tall, and built large at that height. Then my son reminds me when I asked what it looked like, he says, [you know, like that thing xxxx and I saw about a year ago up between the sheds one night.] You see, somewhere about a year ago, he and a friend were out running around with flashlights and a paintball gun after dark, when they see what they thought was a stray dog laying in some tall grass between two of our outdoor sheds. They decide to shine the light on it and then shoot a couple paintballs at it when "it" starts to get up revealing that it isn't a dog, just some LARGE hairy...??? bigger than a man... but "we don't know what it was... so we just ran" kind of thing. They never saw a "face", but did see a large eye that reflected orangish red back at them when the flashlight hit it as it seemed to turn it's head slightly toward them. When they DID see the eye reflecting back and realized that was a head, they observed there was no muzzle of any sort. That's when they realized it wasn't a dog. They didn't stick around for "it" to get up all the way off the ground. They could see that as it was getting up, it was large, hairy, more like a person and not a dog. So he really felt like this Oct. sighting was the same thing they had seen some 6 or more months ago.

So... that's basicly it. After much grilling on my part, I do believe them. I didn't see it myself. I wish I had the presence of mind that morning to grab a gun and stand on the porch and see if something did make it's way up the hill in the little woods. Why I didn't, I'm not sure. Too late to speculate.. only thing to do now is plan for "the next time".

ALSO NOTICED: read complete story

OTHER WITNESSES: 3. My son and two friends. They were coming home from school homecoming game and being over at a friend's house after it.

OTHER STORIES: noted in story

TIME AND CONDITIONS: a little after 1:am Saturday morning Oct.11 / clear out with full moon directly overhead

ENVIRONMENT: wet lake yard, between house trailer and lake.. mostly open with tall grass and marshy area...

++++++++++++++++++++++++++
++++++++++++++++++++++++++

Tuesday, October 06, 1981

White River Encounters: Area Residents See "Something Big and Hairy"

By Doug Caroll

The Valley Advance

Jack Lankford is an avid fisherman and hunter who says he never left a fire unattended until the night of Aug. 22. That's the night he saw a "creature" while fishing in the White River bottom land about six miles south of Highway 50.

Roger and Barbara Crabtree say they live in fear of a "hairy creature" they have seen twice near their Decker Chapel home in southern Knox County close to the White River.

Terry and Mary Harper haven't seen anything, but something attacked their house at 2002 South 15th Street, Vincennes. The unknown assailant ripped and apparently chewed on aluminum siding and tore away part of the metal trim around the backdoor of the house. It left behind teeth marks, blood and tufts of white hair about two inches long.

So far the incidents are unrelated. No evidence of a creature has been found in the areas where the sightings took place. However, Lankford says what he saw was no bear, and the Crabtrees know that people may not believe, but their fear is "very real."

Lankford anticipated a "good bit of fishing" last Aug. 22 when he went to his favorite spot on the lower part of what is called Beaver Dam in eastern Knox County. The fisherman had built a campfire a few yards from the bank and was using a lantern to watch his lines.

The Washington, Ind., resident had been there a couple of hours when he started having an "eerie feeling" that someone was watching. About 20 minutes later Lankford looked up and saw two eyes, each about one-inch in diameter, glowing red from the lantern and nearby campfire glow and staring at him from about 50 yards away.

Lankford could see a hairy body sticking about four feet out of the water, but the light was too dim to reveal the face, he said.

Lankford said the creature looked like a well-build, big-boned man with "extra" long forearms and covered with brown, matted hair. It apparently was standing in about four feet of water.

"It just stared at me and me at it. It was trying to figure out if I was looking at what I was seeing," he said.

The "booger," as Lankford's grandmother called it, appeared to study Lankford, tilting its head from side to side and making no noise, he said. After a short time, the creature turned away, reached to grab a tree limb, and pulled itself from the water.

As it walked away Lankford noticed that the arms extended to around the knees and that it had to weigh "well over 200 pounds."

"It made a loud squeal or high-pitch shriek when it left, something like a young pig would make when you try to hold on to it."

Lankford heard the sound again while he was hurriedly packing his fishing gear. He says he has heard the noise in that area three or four times since early spring, but didn't think much of it.

Since seeing the creature, Lankford has not heard the noise. He said he would like to meet it again.

"The last time I didn't think to follow it because it didn't show any sign of wanting to harm me. I'm one person who respects other persons and beings, and I would like to see the creature captured unharmed and studied," Lankford commented.

Lankford told only his family immediately after seeing the creature. He decided to report the incident to the Daviess County Sheriff's Department after reading a newspaper article about the attack on the Harper house.

"I've talked to people who live in the area, and they said if it is someone trying to pull a hoax they are taking a big risk of getting shot. The sheriff's deputy told me the same thing," Lankford noted.

Terry and Mary Harper, their children and neighbors did not hear anything out of the ordinary between midnight and 6:30 a.m. on Aug. 26, but during that time about four or five feet of siding some three feet high was ripped and chewed, along with metal trim around the backdoor. One piece shows what looks like a claw mark.

"We had the house fans on all night and they can be noisy. We really didn't hear anything," she said.

Terry Harper was leaving for work when he saw the damaged siding. The damage amounted to about $500, Harper said, and included blood, large teeth marks and white hair. Blood was also found near the back light about six to seven feet above the ground, Mrs. Harper said.

The dog refused to come out of its house and had its paws over its eyes and whined when it was checked.

Officials from the Knox County Sheriff's Department have told the Harpers that tests on the blood reveal that it is not human, and that a wolf or some other wild animal may have done the damage. Investigating officers told Mrs. Harper that hair taken from the scene has been lost.

"We don't know what to be frightened of, and I can't say that it is a 'bigfoot' or not," Mrs. Harper said.

Harold Allison, an area naturalist and writer of a weekly nature column in The Valley Advance, studied pictures of the damage and believes no animal native to the area could have caused the damage.

"The only animal I can think of from my experience capable of that kind of damage would be a wolverine. But there are no wolverines within 500 miles of this part of Indiana," Allison commented.

The incident has kept the Harpers busy on the telephone, talking with newspaper, television and radio reporters about the "house attack."

Mrs. Harper has been interviewed by radio stations from as far as Boston, Chicago, Dallas and Los Angeles. The incident received a brief mention on the ABC-TV World News program.

Through a United Press International news story, an investigator from S.I.T.U. Research Services, a private company in Little Silver, N.J., has contacted the Harpers and currently is looking into the incident.

"The investigator thinks it's a big foot, but he can't be sure because we didn't have any blood stains left to send him. He said if we could get him a blood sample, he could tell us exactly what it was," Mrs. Harper related.

S.I.T.U., which reportedly specializes in unexplained phenomena investigations, sent the family a report of a 1977 attack in New Jersey.

The New Jersey incident involved a creature like the one described by Lankford, but with a human face covered by a beard and mustache. Wood panels on a barn were ripped up and chewed at about the same height as the Harpers' house.

The most recent sighting of what one area newspaper has called the "Knoxness monster," occurred Sept. 26 at about 2:30 a.m., along the Decker Chapel Road, west of Highway 41.

Crabtree was returning with his family from Princeton and was less than two miles from home when he saw "something big" walking in the road.

As Crabtree came closer he noticed fur, long arms and a "skipping walk like an ape." The headlights appeared to startle it, Crabtree recalls, and the creature swung its arm at the car. Crabtree swerved off and back onto the road to miss the creature and stopped to watch as it continued its walk down the road.

Crabtree's wife, Barbara, who was awakened by the quick turn, persuaded her husband not to follow and to call the Knox County Sheriff's Department.

Mrs. Crabtree said she had seen it the day before in a cornfield near the family's backyard, a "dirty, white-haired creature" not more than 50 feet away.

Mrs. Crabtree grabbed her two pre-school daughters and backed to the front porch, she said. The creature "took a couple of steps" toward her but stepped back when the family dog started barking and ran toward it.

She got her daughters and nine-year-old son, who was throwing rocks at it, into the house and locked all doors and windows. She tried to call the sheriff but was unable to get through because of a busy party telephone line, she said.

In her view the creature was about seven to eight feet tall and weighed around 500 pounds. It was covered with "fuzzy" dirty white hair except for its head, which was brown hair.

"It had a pinkish face and big, glassy eyes. The thing had an awful, sour smell, something like dead meat that had set out for three or four days," she said.

The creature also made a growling noise, which the family has heard at least two times since the second sighting, Crabtree said.

The sheriff's deputies have been unable to find any evidence of the creature and consider the case closed, officer Jim Wilson said.

"The department is treating it as an unconfirmed sighting because the Crabtrees were the only ones to report it," Wilson explained.

The family is now looking for another house and has purchased a shotgun.

"I don't care what anyone thinks. I saw what I saw and no one has to believe me," Roger began. "When nightfall comes around here, my family is plenty scared. I don't even go out after dark."

The sightings and descriptions by the Crabtrees and Lankford are similar to reports that Allison has investigated in the past three years. Allison, who has looked into about 25 sightings, began getting reports from Knox County only recently.

Numerous reports have come from near Shoals along the White River in Martin County.

"Their (Crabtrees') descriptions tally up with the others, and I feel they are sincere after talking with them. There is something out there, but I have no idea what it is," Allison said.

Allison added that if such a creature is alive, he hopes it will be captures and studied and not hunted down and killed.

Bibliographical Information:

Vol. 18, No. 6, Vincennes, Ind.,

++++++++++++++++++++++++++
++++++++++++++++++++++++++

YEAR: 2004

SEASON: Spring

MONTH: April

DATE: 2

STATE: Indiana

COUNTY: Kosciusko County

LOCATION DETAILS: 300 feet east of the intersection of county roads 1100 north and 300 east. This is east of Waubee Lake in Milford, IN.

NEAREST TOWN: Milford

NEAREST ROAD: County roads 1100 N & 300 E.

OBSERVED: On Friday, April 2, 2004, at about 5:00 PM my wife was returning by herself from a trip to drop off our daughter at a friend`s house west of Waubee Lake in Milford, Indiana. She was travelling east on County road 1100 North. She approached the intersection with County road 300 East, which leads into Oswego, and came to a stop at the stop sign. She looked left, right and briefly ahead. Before beginning her turn she noticed movement directly ahead on road 1100 N. She saw crossing the road from left to right,(heading south) approximately 250 feet ahead, two large, black, hair covered figures, taller than an average man walking on two legs at a hurried pace.

Her first reaction was to say out loud "what the heck was that?". The way they were walking very close together and hurrying suggested that they were aware they were exposed in the open and wanted to get back to cover quickly. By the time she saw them they were in the middle of the road but she saw them take 5 to 7 steps over several seconds. They were stooped forward and looking down and swinging their long arms quickly. After this they entered the woods. She said they were close enough to have scared her if they had looked her way. Also, they were close enough to see they were not just men in black clothing as there were no divisions where shirt, pants and a hat would be; only solid black from head to toe.

My wife, who is a college educated professional, then made her turn onto 300E thinking she would then have something interesting to tell me immediately upon arriving home. When she reached home however, something else unusual happened. Instead of telling me, she apparently blocked out or repressed the experience from her memory, as if in denial. I have been a Bigfoot enthusiast for many years and have read of this happening before to people seeing a Bigfoot. It is like your mind refuses to believe what the eyes are telling it because it is so new, different and without a reference point. About 2 days later we were talking and I said something that brought it all back to her suddenly. She then told me the whole experience in detail, showing signs of alarm and even imitating the way they walked. Of course I wasn`t about to believe her without asking questions. After all I was the Biology major in college, I'm the outdoorsman and I'm the bigfoot enthusiast - I'm the one who should have seen them!

I asked all the typical questions that the BFRO interviewers asked and she answered them all correctly, even adding additional detail. I asked if they might possibly be some young high school guys in all black Gothic clothes, as some of them like to wear. The answer was "they were absolutely not men in black clothes". The thought also occurred to me that she was messing with me - playing a joke to mislead me, so I asked. This was the last straw; she got hurt and angry with me for not trusting her after 27 years of marriage, and I got the silent treatment for 2 days and had to do a lot of apologizing. I'll have to just agree with her - based on the evidence, she saw 2 Bigfoots crossing the road. Wish I had seen them!

Hopefully, by learning as much as I can about this creature I will be able to remain calm enough to stay and observe it, if I ever see one, instead of freak out and run as most do. Some may say "your wife learned all this from listening to you go on and on about what you read on the internet. Believe me, when she tells her story you can tell it was a very real experience for her; whether anyone believes her or not she knows what she saw, and will say so.

ALSO NOTICED: Nothing

OTHER WITNESSES: None

OTHER STORIES: Have heard "possible vocalizations/screams" when camping south eastern Indiana

TIME AND CONDITIONS: Clear

ENVIRONMENT: Low, wet looking area on north side and wooded on south side of road

----

Follow-up investigation report by BFRO Investigator Stan Courtney:

I spoke at length with the witness by phone. She described the animals as being somewhat taller than six feet tall. They walked kind of humped over and in a hurry.

+++

YEAR: 2002

SEASON: Spring

MONTH: May

DATE: 11

STATE: Indiana

COUNTY: Johnson County

LOCATION DETAILS: [edited out by request]

NEAREST TOWN: Nineveh

NEAREST ROAD: Old Nineveh Road

OBSERVED: A friend of mine cornered me late one day and told me a strange tale. He was going to shoot a rifle that he owns and he said he smelled something dead or rotten. He thought maybe someone had killed a deer or another critter. As he proceeded to walk through the woods, he said he saw a huge animal hunkered staring at him. He said the shoulders were massive. His rifle was loaded, but he felt he didn't have a big enough gun (.223 cal.) They stared at each other for a few seconds, but my friend turned to the right and proceeded forward to get away from the "MONSTER" (his word). He said it bolted in the same direction as he did and ran past him in a second and a half. I said was it a bear? He said no. Was it on two legs or four? He said two. It broke tree branches as it ran thru the trees and brush. He was so scared that he did not return to the woods for 8 or 9 months, and when he did, it was with a loaded 12 gauge shotgun loaded with deer slugs.

The first person he told was his wife and she laughed at him. I listened to his story and did not laugh at him because he was so serious. I will say this, he didn't know about the bigfoot sites or even mention the name bigfoot during the telling of this tale. One other thing, he was carrying a bag of apples, he felt later that the MONSTER was after that. That part of the story I know is true. I saw the bag!

I read your reports, and can tell you I don't know whether to believe them or not. I think my friend saw something strange and was scared. He warned me not to go into those woods any more. I said bigfoot didn't eat you so he won't eat me.

He said he wanted to post a sign next to the woods that said "MONSTER IN THE WOODS", but I said no because I didn't need every guy in Indy with a 300 magnun chasing a myth or legend. I said if you did see a bigfoot, your one lucky guy! This is the truth. No hoax. Thanks

ALSO NOTICED: I have seen broken branches 10 to 12 feet up on young strong trees. Twisted is the word. I have heard whoops and strange animal sounds. But it is the forest!

OTHER WITNESSES: One

OTHER STORIES: Yes, but will report them later.

TIME AND CONDITIONS: 10 a.m.

ENVIRONMENT: Wooded and mixed with farm fields, small hills and ravines.

Follow-up investigation report by BFRO Investigator Stan Courtney:

I spoke with both the main witness and the person who submitted the report and can add the following information:

• The witness said the animal was a female because he could see large breasts.

• He stated the animal was at least 9 feet tall and weighed 800 pounds. He was adamant about how huge the animal was.

- The animal was very stocky and fit looking.

- Its hair was blackish gray and about 2 inches long. The hair was not matted at all.

- Its neck appeared very thick and the arms did seem somewhat longer than a human's.

- The closest he was to the animal was 20 feet.

- He could not see the face until the animal was several hundred feet away and he could not make out any facial details.

• The person who submitted the report hunts in the same woods and has heard what he considered to be whooping sounds at about half a mile away.

++++++++++++++++++++++++
++++++++++++++++++++++++

YEAR: 1999

SEASON: Fall

MONTH: October

DATE: around 23rd

STATE: Indiana

COUNTY: Jefferson County

LOCATION DETAILS: Near Jefferson Proving Grounds.

OBSERVED: I am writing to report a sighting while I was deer hunting. I have been hunting since I was 10 years old and have been hunting by myself since I was 13.

My friend and I were picked for a special bow deer hunt in southern Indiana, Jefferson Proving Grounds. I had set up my tree stand along a fire break where I could see both, the opening and the wood line ahead of me. I had observed several deer around my stand about 12:30 that day and I took a shot at one of them. I tracked the deer across the fire break and into the woods where I lost his trail because of the dense briars and rose bushes. I could hardly walk through the stuff so after about 30 minutes I returned to my stand. Throughout the afternoon I watched the air force do practice runs over a restricted bombing area not far from where I was hunting. Later that evening when things had finally quieted down I heard some howling off in the distance. I figured that it was a pack of coyotes. About 5 minutes after they quieted down I began hearing some noise coming toward me from the deep woods. I thought that it was a deer running from the coyotes. I first saw it about 50 yards out coming toward me. It was about dark and as any hunter knows, thats the best time to see and shot a deer. I was ready and watching for an opening. When I had my first opening to get a shot I noticed that it was not a deer, but appeared to be a hunter. I watched as the figure ran across the opening through the brush to a second opening and then to a third opening where it crossed the fire break and then back into the woods behind me. I thought that it was odd because I have never seen a hunter dumb enough to run through the wood during the prime hunting time. What was even more scary was the fact that it had taken me 10 minutes to walk through the same brush that this thing had run through in 20 seconds. It was completely black from head to toe and had a hump shape on its back. It stood about 6 foot 6 inches. The reason I can say that is because im 6'1" and this thing was bigger than I am. I waited in my stand for a few minutes and then I got down and walked "fast" back to my truck. I met my buddy who was over the hill about 500 yards and he said that he heard whatever it was going through the brush but never saw it. I told him that it was a Big Foot and he laughted at me until he could see how serious I was about the thing. We asked at the checkout center if anyone else was hunting in the area and they said that we were the only hunters in that area and the two areas around that.

OTHER WITNESSES: Hunting partner heard it as it ran through the woods after crossing the field, but did not see it.

TIME AND CONDITIONS: Approaching dark, but still light enough to hunt.

ENVIRONMENT: Open Field, with pockets of brush, forest to the rear of the witness.

++++++++++++++++++++++++++
++++++++++++++++++++++++++

YEAR: 1991

SEASON: Summer

STATE: Indiana

COUNTY: Harrison County

NEAREST TOWN: Corydon

OBSERVED: My name is Don. I lived with my brother in a small A-frame cabin on Corydon-Ridge road in Corydon In.47112. It was summer of 91 ,it started with hearing sounds"howl-screams" in the woods behind the house. It scared the heck out of us and our dogs but we blew it off until it happened again. Being young and very stupid at the time (I was 20) I got the brilliant idea of shooting into the woods from where the sounds came from. From that night till we moved out about 2 weeks later we would hear the screams getting closer We kept shooting up in the air and out into the woods about 200 yards away the screams would stop after 5-10 shots but they would start a night or 2 later. My friend James came over and camped back in the woods. He was only there for several hours he said he saw something humanoid/hairy/tall walk between 2 trees He left shortly there after. The next night my brother woke up to a large shape staring at him in the window it had to be at least 8 foot tall. The next night my friend Mike stayed over out of disbelief. That's when it howled it sounded like it was in the house it was so loud. It was right under my window(my bedroom was upstairs).We went out the next day no tracks but the back steps(WOOD 2x8's)two of them had been caved in from something heavy enough to break the door jam. The knob was not hurt the door lock was still locked it was pushed with so much force it splintered the frame where the bolt met the wood. My dogs were cowering upstairs they messed all over the upstairs they wouldn't even come to me. The house was not touched inside. It smelled like feces/skunky/wet hair/urine/ it is hard to describe it wasn't overpowering it just was strong .I moved out the next day. No one would believe me, so I've only told a few people. I asked the local people about it they all said I was nuts or they would slam the door in my face. Some people bought the place a few years later; it was rental when I was there, I thought it was odd they disassembled the tiny cabin and hauled off on a flat bed truck ,they sold the land. There is a large brick home there today. Corydon is in Indiana,Harrison county,USA 47112

If you want to ask any questions contact me I never saw anything I guess this would be a bigfoot I don't know what else it could be.

ENVIRONMENT: Woods near a forestry, a lot of sink holes and caves.

++++++++++++++++++++++++
++++++++++++++++++++++++

YEAR: 1998

SEASON: Spring

MONTH: April

STATE: Indiana

COUNTY: Harrison County

LOCATION DETAILS: take I 64 to exit 121. take route 131 south for about 7 miles. Turn left onto wiseman road. At corner of wiseman and union chapel is my old place.

NEAREST TOWN: Corydon

NEAREST ROAD: Wiseman/Union Chapel

OBSERVED: Me and my roomate were watching a movie, when we heard this ungodly sound from out side. It was this wierd wail/howl/scream. It started out real low and guttural, like a cheap frankenstein groan, and then shot up the octave range to an amazing high pitch wavery howl. It sounded like a cross between a woman screaming and a fire siren. It got quiet for a mintue and the the screaming started up again, but it sound like it had moved closer. All the dogs in the neighborhood went nuts. They sounded like they were dying or something. After the second wail stoppped I worked up the nerve to open the door and go outside. It smelled horrible outside, and I got the distinct impression I was being watched. So, I went back inside, and hoped that what ever it was, it was gone.

ALSO NOTICED: I worked third shift, and sometimes there was a god awful smell in the air when I got home from work. Not the farm smells either. This was a nasty musky, rotten smell that seemed to stick to your skin.

OTHER WITNESSES: My old roomate Darrell. We were watching a movie. I think it was the prophecy. Plus any nieghbors that heard it.

OTHER STORIES: Heard that a large hairy thing had been seen in the area before. Didn't believe it till I heard that noise.

TIME AND CONDITIONS: It was about 1 in the morning, and it had just rained a while ago, but was a clear and starry night at the time

ENVIRONMENT: Farms, Heavily wooded in spots, lots of sink holes.

Follow-up investigation report by BFRO Investigator Tony Gerard:

I spoke with the witness by phone. He said there were two or three vocalizations total lasting about seven seconds each time.

Prior to the vocalizations he had experienced the odor on several occasions during the Spring, and possibly one time after hearing the vocalization.

++++++++++++++++++++++++
++++++++++++++++++++++++

YEAR: 1973

SEASON: Summer

MONTH: July

DATE: 10

STATE: Indiana

COUNTY: Hamilton County

LOCATION DETAILS: Just outside Westfield on Carey Road.

NEAREST TOWN: Westfield

NEAREST ROAD: Carey Road

OBSERVED: (Report as stated in telephone interview with investigator)When I was thirteen years of age a friend and I were riding our bicycles on a country road near some abandoned railroad tracks.

We were startled to hear a maple tree come crashing to the ground. My friend immediately left on his bicycle.

The tree was about two inches in diamater. Looking up I was amazed to see a very tall large animal standing next to the tree. It was about seven and one half feet tall and very hairy. I would guess it weighed about four hundred fifty pounds. I was about ninety feet from it. I froze in disbelief and we stared at each other for at least two minutes. It then slowly started walking towards me and I panicked. I turned my bicycle around and road home as fast as I could go.

ALSO NOTICED: None

OTHER WITNESSES: Just one, but he did not stick around.

OTHER STORIES: None

TIME AND CONDITIONS: 1 p.m.

Clear, dry and hot.

ENVIRONMENT: Woods.

Follow-up investigation report by BFRO Investigator Stan Courtney:

I spoke with the witness by phone. Since he no longer has internet service I filled out the sighting report for him.

He stated he was unable to see any facial features because the animal was too hairy. The animal's arms were very long and as it walked towards him it was on two feet.

+++++++++++++++++++++++++
+++++++++++++++++++++++++

There are several other Indiana reports, but to make the book interesting, we will make a volume two for those needs, but as for now, Steve Abney hat lives in Indiana and has researched Indiana for seeral years has moved on past the flesh and blood bigfoot, and has experienced the more advanced segment of the bigfoot phenomenon, and ha worked closely with one clan in habituation, and here is that story for the reader to ponder on, it is a fully true story and not some imagination, or craziness,

This book details our sightings and encounters of the Hairy Folk known as Sasquatch.

There is a table of contents to follow in numerical order with the dates and/or times, and to the best of our ability, as well all conversations with Kochi (pronounced Kauchee), it is our hope that you enjoy reading the book as well as we did writing it. This is a unique story never heard before in American History. We had to change our life to suit the Hairy Folks in order for them to communicate.

We dedicate this book to all those that have died at the hands of terrorists, since 9-11-01 until to date.

With the things that are going on in the world today, Violet and I firmly believe we are living in the last prophecy of our lord and savior, Jesus Christ. So, to the story.

TABLE OF CONTENTS

1. Message from the forest people
2. Just living to be comfortable
3. A Living Culture
4. Smiling and joking all the way
5. Important, but Humble

6. Change of scenery

PART TWO, CHAPTER ONE, MEANDERING THINGS THAT BELONG.

PART TWO, CHAPTER TWO, IT IS DESTINY.

CHAPTER ONE

MESSAGE FROM THE FOREST PEOPLE

Message from the forest people:

Change the way you think, humans, before it is too late to turn back. The destruction that you have wrought upon the earth, our home, is too important to destroy. We all live here, the two legged and the four legged.

We have had to learn to survive the ice ages, and you have not. We have had to learn to survive the floods that God sent to destroy humanity, and you got worse. You humans never learn from your mistakes. Turn to your god and repent before it is too late.

Gifts are the way we survive; gifts left to us throughout the ages from our ancients that were given to them in the Garden.
You humans used to have them but you abused them, and they were taken away after the great flood. You were given another chance through the 8 that survived. It all started out fine till time crossed over to evil again. We hid because we saw the evil in you. After a time went by, the earth started to shrink in our homeland, the forest and mountains and river valleys. Then we showed ourselves.

As Violet and I understood things, we were chosen by Kochi to undertake this great and sacred message to the world in writing this book. Many things you will not understand till you open your heart, open your mind, and lose the total fear of what is really in the world that humans cannot see or hear. For you see, we were the same way till Kochi interceded.

There are many things we have experienced that have never been written before, and in this book, you the reader will open your eyes and question us many times over as we take you through the message that the forest people want you to understand.

In the beginning, when Kochi froze our minds through mind speak and warned us of the red rogue, he left the gift of understanding in our minds, as well other gifts that you will see ,**what we are talking about** as we take you through this journey.

So first, let me take you through the compassion of Kochi. While I was standing there frozen at midnight holding and hugging my wife, scared to death that we were going to die that night, the energy we felt was changing to warming of the heart, a heart of energy that showed us as sinners in a world gone mad. And the energy went to the changing of the heart to an overflowing of joy. Just being among the forest people as Kochi was showing us we were surrounded by 14 others that thought the way he did, that humans were worth saving, and the bravery we showed hugging and holding on to each other in the fierceness of his mind control, Kochi's heart and energy melted into us. We felt his compassion for us, and he showed us his face. You could see the pity and the sadness and the total compassion in his heart. He was a true leader of his clan, and we knew from that instant that something unique was going to happen to us, but we did not know what.

The next thing we saw was the fierceness of a true giant warrior of ancient times, a time when giants walked the earth and ruled with an iron fist. Then and there we knew that Kochi was that type of giant, but for a few differences. He was huge, yes, but he was a teddy bear of a giant until battle was called upon him, and then he would rise to the occasion.

Then, we felt the energy change again, and we saw and felt the deep sadness for his kind in this world of love and hate and confusion.

We saw through his mind the way they lived and that they were happy and free. Then that all changed, and we saw the way they had to skip through the changes of humans around them in order to survive, and it saddened them. Kochi show us the rich and all-changing forest around them in the old days until now, and what we saw was a terrible wreck upon the earth. This is what they had to live through in today's world with humans. To help you understand, here it is for you.

Let's say they walked through the forest for many years and were always happy with the way their home looked and felt, and they knew that they were safe. Then one day they walked through the forest and came upon a clearing that was being bulldozed by humans. They watched from a distance as the diesel fumes were belching out of a metal exhaust pipe, and they smelled the noxious fumes and watched as the living trees around them were being destroyed. This was a sacred place for them, this was their home and their graveyard, and they could do nothing about it.

Now I hope you understand the sadness this caused them, and that it makes you become aware that we, as humans, are destroying the homeland of others that do not want to be known or bothered in any way. This is what we saw and felt that very first night, and we had to hide this feeling, we what felt and saw, until the right time to write it down for others. Kochi has come into our world many times, correcting us as we go, and believe me when I tell you, we made many human mistakes.

It is time to move on to other lessons of the way they live, but they wanted us to write what we saw and felt throughout that first night as we were hugging and holding each other in fright.

CHAPTER TWOJUST LIVING TO BE COMFORTABLE.

While most people or researchers would cower in fright at the mere sight of the forest people, the forest people would feel the same way if they were allowed to be seen on a daily basis, being free to walk among us shoulder to shoulder. But the real secret is, in fact, they do that all the time, only secretly, under the radar so to speak, because of mind control or using EMF to put a shield over the clan to protect them from being seen, so they can go about their daily business and not be bothered by mere humans with all the negative thoughts and actions.

But on the bright side, they are happy to play a joke on the humans at last, so they can do what they were put on earth to do, and that is to protect the forest and guide humans on the right path.

To live comfortably in their culture and atmosphere and not to be bothered, in this secret scenario, the clan leader is the busiest of them all, running from place to place on private property, checking zones and his members to be sure they are safe and secure.

The leader's responsibility includes checking the people that are the watchers of their territory and making sure the mates of his members are well-fed, comfortable, and safe, then organizing food gatherers and hunters for the food they will need.

For instance, when I was fishing late one night, and our sister-in-law came from Kirk's house to tell me it was supper time, Kochi was standing on the border right inside of the overgrown Christmas trees, and my sister-in-law walked right past him and two others. He did not walk out until she walked past him again on the way back to the house. Remember, there was nothing there to be seen, but as soon as she left, I was compelled to pick up the camera and take a picture. They were right in front of the lens, invisible. He walked casually out onto the driveway and walked to his secret hiding place with the two others following. He was calm and safe to do so in plain sight.

The very idea that you walked past a twelve-foot being and knowing about it would have frightened the hell out of her. That is why Kochi was invisible. Not only do they do this on private property, they do this in the wild as well. But understand one thing, some clans do not have this gift to hide so they have to use other means to become allusive.

Comfort is respect to them, and they will continue to fool humans as long as they remain comfortable in their environment.

Respect, that is the key word in their world. When they feel safe to communicate with humans, they will remain hidden and mindspeak in order to remain safe. Yet they show respect to you by not scaring you with their looks, because they know they scare humans to death. They calmly mindspeak to you about certain things, but there will be times when, like they did to us, they give you a warning about certain things. It is imperative for you to listen to them and do exactly what they say. They will not want you harmed or use someone else to relay that message to you. Please be heedful.

There will be a time when this entire scenario will save your life like it did with ours. Being able to hide like they do, helps them to hunt, to teach their children to live in their culture in a safe way, and to continue to grow with the safety of the clan.

TRAVEL

When a young juvenile is at a certain age, he is free to travel to other clans to learn and further his education about adulthood. When he leaves his own clan that has the safety factor of cloaking, then he takes the risk of being seen outside of the shield that covers a certain area. They mostly travel at night and hide or forage deep in the forest during daylight hours.

When the clan leader or anyone in the clan has to travel, they take the same risk as well. Traveling in a family unit is the most risk of all. Usually, they take the shield with them at all times when the clan moves from one area to another.

For instance, daytime in the country in the **flat ground of** Indiana with rivers and surrounding forest in certain areas, with Indian mounds, natural lakes, and Indian ceremonial **battlegrounds**, as well as grave sites, require them to be very cautious. Southern Indiana is much different with its ravines, hills, and thick natural forests. So, let's take the scenario that they want to travel from southern Indiana to the Wabash Valley north of them. They will extend the shield two miles in front of them. They begin by traveling as a family unit under the shield, as not to be seen, and they stick to the forest as much as possible. Then besides roads, they stay as far as possible away from the roadway and stay in the low spots in the field, and other times they hide all day and just travel at night. Getting around large cities, they circle them on **waterways** at night. Can you imagine a family unit of forest people traveling through a city invisible?

Well, they travel around it and find ways to go over the roads through bridges by going under them and continue to travel north. There are plenty of forest thickets along the rivers and creeks. During the day, when they are forced to go through a bean field or corn field, they know that they could be seen, but the respect and the fellowship of the clan leader are their safety ticket. And the trust they have is stability for them. Finally, they end up north in a heavy forest environment. This is the way they travel safely.

SECURITY

This is the number one rule of all clans, no matter what state they are in and no matter where they settle after they travel. There are guardians to the clan like a secret army of watchers. They signal through tree knocks and screams at night and correlate travel plans through the clan leader. Then there are the guardians that secretly travel, surrounding members when they are on a mission for the clan. Many humans see them as tree-peekers. Some humans say they feel surrounded by something watching them but cannot see them. They do not know it, but the humans are right. They always surround the person until they find out their intentions. Then they move on.

Some individuals will cloak and walk among humans as a test, or because they are not afraid. Being brave is something the red man did many times in the past. It was called coup. The safety of the clan is vital to their survival. At times they mark the trail they travel by leaving markers for stragglers to find them.

GATHERERS AND HUNTERS

Being comfortable in your surroundings is very important to humans and to the forest people alike. They do many things that humans did when humans lived in the stone age before the white man came and destroyed the habitat. These relic hominids were and still are the boss of the woods, as the Native Americans has always said. The Native Americans copied their way of life and lived near them for many years. The legends are proved through DNA. As we talk about this, remember the natives of our great land have survived thousands of years of living in the wild before the white man came to this country. As gatherers and hunters, there are no equals, and some still live that way, too, through their traditional ways.

They let the women of the clan gather at night through the safety of the guardians, while other women or juveniles watch the children. Sometimes there are certain teachers that go out at night with the children under the guardianship of the guardians and teach them how to fish, hunt, and make shelters and territory markers.

Gathering flowers and leaves, bark and moss for food, and medicines for the clan is the responsibility of the clan as a whole. The safety of women and children is so strict that if a human tries to hunt them, that human would be killed and would disappear from the face of the earth, with no trace, and that would go for predators of any kind. This is the safety of the clan women, and at no time will the women be allowed to be seen in the daylight without the express permission of the clan leader.

ELDERS

Wisdom is the key to survival for any and all clans of the forest people, and that wisdom comes through the elders and their great wisdom of survival. Elders can live alone a little distance away from the clan. They can run the clan when the clan leaders are away but are always involved with the business of the clan as a whole. Wisdom comes from many years of living in the environment of the clan in different areas of the country and always surviving those winters and the things that happen unexpectedly.

They can handle the extremes that come along and

Make life and death decisions through those years of wisdom.

++++++++++++++++++++++++++++++++

WARRIORS

You would not think that a clan of forest people would need warriors, but that can happen. When other clans start to run out of space or want to take over a rich area where the clan lives, the clan will come together and fight the ones trying to take over, and they always follow the clan leader. The ones guarding the women and children are considered warriors.

CHAPTER THREE

A LIVING CULTURE

Many people would never know that other cultures exist in the world. Because of all the studies of the world culture, science thinks they know all the cultures of the world, but in fact, we learn of new ones every day. The forest people have about the same culture to survive in the wild from clan to clan, but many times it differs from clan to clan. Their culture can be complicated or simple, it just depends on the way you look at it. Their safety and security demands certain things must be done in certain ways.

Kochi's words to us and watching how they do things around them, we have a small idea, and working with other clans, in other states has brought us to the moment of writing this book. The culture of the forest people will be a simple thing for those who understand primary survival skills in the wild, but to others it can be difficult. So to the task at hand, when Kochi gives advice or even a command, it had better be done right this instant. To explain this, they do not mince words. When one says, "You are mine," they attack. It's just that simple. Words are said, then the task is done NOW. Period, end of conversation. That is why I always say, when one enters your mind and asks a question, you better be prepared to tell the truth and right this instant, your life can depend on it. Or, if they tell you about their clan, then you take the words to the bank.

Tree shaking is a warning to humans, but the thought is "Get out now." The hesitating is done for one reason, and that reason is in fact, you are human and not a clan member. They are telling you, they understand, if you do not leave right now, they do not want to hurt a human, because that would bring harm to the clan later. But if it was a clan member, then leave right now had better be done now. Or else. I have taught over and over that everything they do is based on one rule and only on one rule, safety and security of the clan, no matter what.

For instance, we ran across one clan that had a visitor and who was accepted into the clan for an extended time. The big guest wanted the clan leader's mate. When the clan leader was a few feet away, the guest charged the clan leader's mate and tried to drag her off to be his. The clan leader and others of the clan charged the guest with sticks and beat him off the mate, and banned him from the area. It was done immediately.

We ran across a clan that needed our help in getting rid of another clan that was invading their territory and stealing mates and other things. We set up a sacred circle of stones and prayed in the circle. We screamed at the other invading clan. This brought all of the clan that lived there together, and they stood with us humans to stop them. The invading clan knew we were working with that clan and would use deadly force if we necessary to protect the clan we interacted with. This brings not only harmony, but deep respect, and shows you are not afraid to put yourself in harm's way to love and respect them.
 This brings them closer to you. Culture is so important. Just remember what I said, safety and security mean everything to them, and you provided it to them. It is scary as hell to put your life on the line, but that is what Christ did for us.

There are times when you can ask a serious question of them, and if that trust level is there, then you will get a direct answer. Other times, you will get a report back, just saying to you, never mind, that is not important at this time. Here is an example of Kochi talking to us:

Kochi== Why do you smoke that nasty weed?

Steve== I like the calmness it gives me, and it tastes good.

Kochi== It messes your mind up and causes trouble to your body. Never mind, you will not listen, end of conversation,

Other times we ask a question of him, and we may not get an answer in mind speak, but that night he will take your mind and through his eyes, he takes us on a trip through the dark and green dangerous forest. Through parables, he answers the question we ask, much like a live reality show. I may be safe in bed but in someone else's mind. Scary? Yes, it is. But after a while, you can see this is the best way to teach. Jesus taught in parables as well, and the simple minded learned that way. So, the scary can be quite pleasant after a while,

There were a couple of times when Kochi did this. He took me through a tunnel of turning green mess, like a portal, only through his eyes. At the end of the tunnel, the most beautiful area came into view, and through his eyes you could feel the true freedom and beauty of the living forest. As long as you knew how to survive in this new and fantastic world, I loved to do that through his eyes only. I was not prepared to go it alone, lol, no way. I knew I could not survive. But think how God describes heaven. There are dimensions of clear reason and no killing fields, but only where all energies converge in love and harmony.

Living in perfect harmony on earth with their own clan is still not perfect, but as close as it comes. Remember, whatever positive comes forth, there is a negative ready somewhere else to rear its head. I do not bash anyone nor sarcastically say anything about the forest people, but rather walk away from the naysayers. It's the only way to keep the truth in your own mind and heart. But as Kochi said one day, "You have your god. Listen, and do."

Fierceness in battle is good and bad; good because you win, but bad, because the life you fought is gone and buried, and now is the time to forget and move on to the next chapter.

I asked Kochi about those humans that hunt him and his kind, and he was troubled with an answer. He thought about what to tell me for several days. He told me that he was troubled by my
question because there was hatred in his heart for those who knew the forest people but came after them as if they were hunting an animal. Many had been wounded and some crippled for life. His people had to look after them, and he was tired of picking up the pieces. Then he softened his tone and said, "Hatred is a hard feeling. I feel bad about saying that, but dislike is more like it. But what I say is true."

CHAPTER FOUR

SMILING AND JOKING ALL THE WAY

During the construction of this book, I had to write about all the good times we had with Kochi and company, and there were more good times than bad. In fact, there were not too many bad times that I remember, except being scared the first time. All the jokes they played at our expense was something I will always cherish: like the first time I heard them laugh at us, or their first chuckle, or gray boy being scared of me after being seen, and being clumsy and running away, and the beach ball landing at our feet in a seemingly empty forest. There are a lot of amusing things these forest people enjoy doing to humans. Their number one is scaring them and reading their minds during the reaction, throwing soft pebbles, shaking trees in front of you, and stealing food while your back is turned during the daylight.

I believe the best one was when Kochi scared the heck out of the four teenage campers in the hollow one night with the lantern incident. Scaring people is one of their greatest joys.

Time always tells *if* the forest people love you and are willing to step out for you. It seems like everything is spiritual for them. This is a solemn matter for them, and the great patience the forest people show cannot be measured. It is a way of life and death to them. Survival is the measure of existence. Being laid back and at the same time being life alert at all times is a way of life for them. Being a relic hominid means doing and living just that way. There can be a serious side to them, but like humans, it can be boring at times, then again, like humans, the humorous side comes out. There will be a time when humans can learn a lot through the lives of the forest people and the wisdom they use of the ancients.

CHAPTER FIVE

IMPORTANT BUT HUMBLE

Kochi came to me many times to tell me of his travels and what he did as a clan leader and clan elder in the circle of his culture. Here are just a few of those travels and how important this was to the clan and other clans living in other territories around his clan's territory.

Before Kochi became the clan leader, his father before him was the clan leader of the same clan. His father would only help the main clan of his, and Kochi saw that that was wrong. He changed that when his father died of old age. His father died of infection from gum disease. and Kochi took over at the age of 30 years. He had made two trips a year with his father at main clan gatherings and learned the art of persuasion watching his father bring clans together for a common good of the forest people. They prospered under his ideas. Kochi did the same. Aafter his father passed, there were times when Kochi had to take chances and had to make a hard decision in order to help those of his clan, like with the red rogue. The red rogue was a bad clan leader from the north of him in a different county. The red rogue was kicked out of his clan because of bad behavior and aggressive towards humans. He broke the number one rule of the clan, which is always the safety and security of his clan. Kochi needed someone to take care of an elder that was crippled and deaf and was seen all the time walking on the gravel roads at night by humans. He needed someone to hunt for him. The red rogue came into Kochi's area at that time begging for a second chance to prove he was better, so Kochi took the chance and gave him an opportunity to prove himself. Two or three months later, after all the hating and grumbling at humans and then chasing Judy, Kochi had to kill him and bury him deep. It almost took Kochi's breath away to think about it. His fierceness always took over to make humans and his clan safe. He would go from one clan to another when the gathering came close, and he would be the ambassador to all of them at the clan gatherings. He would take three other large forest friends of the clan. One was taller than Kochi by one foot, a 13-footer, but skinny. The other two were 10-footers. They always travel under cloaking conditions as not to be seen.

As I said, Kochi was, and still is, an important part of western Indiana clans, and an elder and a diplomat for other clans. He is busy and travels a lot of the time. When he is gone, there are four 7-footers that help hold the clan together while the four travel to help other clans.

They are the four that fooled us the one night going past our tent, lol. I love these guys, lots of fun out of them. After the red rogue died and was buried, a large grey forest person came from the west with his son, the smaller two-toned grey Bigfoot that we fed. The father, the large grey Bigfoot, came in to help the grey, deaf, and crippled Bigfoot while the grey son stayed in the pine forest part of the overgrown Christmas tree farm. It was here he was having fun with humans and interacted with us, while his father was taking care of the deaf and crippled Bigfoot. Loyalty brought the two, father and son, to Kochi and a need to help others. This is being humble and yet important as well.

Kochi always handled Judy with love and kid gloves. He knew his size scared humans, but at night when Judy came out to the front yard, he was completely visible to her and towered over her. He loved her loyalty and bravery and still pitied her for a bad husband. But he showed so much compassion when he towered over her at night and was fully visible to her. He protected her and left gifts for her. Extra deer meat was left at the back door for them. The husband threw total fits at this. The husband worked at a car sale in a small town nearby. Early one morning while it was dark as dark black tar (Kochi had broken the porch light during the night), the man got into his car and started it. He tried to back out of the driveway. The rear of the car was yanked in the air scaring the man to death. The car then was dropped down and the engine died. When he looked out the back window, he saw huge red eyes twelve feet off the ground glaring at him. Kochi entered the man's mind and told him to be careful how he treated his mate or something bad would happen to him.

Kochi disappeared into the blackened night and left the man shaken from the encounter. He later, after the shock subsided, went into denial and shunned the forest surrounding him. He always denied the truth by shutting his kids and wife out from the reality of the forest people on his property, and this brought Judy to do things in secret with the kids at night while the husband was asleep or cowering under the covers, while the forest people were there to visit. As time went on, Kochi had the bean field fight with the red rogue. It filled the whole area with nightmares and terror, and Kochi decided to move to a new area, as not to hurt humans. Because of the screams and terror of that night, the very thought of humans bringing out guns was a fact that Kochi could not hide, so the decision to move not too far away came, and they seemed to disappear. I knew where he had moved and it was 9 miles northeast as the crow flies. He would use the old area as his hunting grounds and take the food home. In the meantime, Judy got a divorce and moved with her kids. She too disappeared.

As far as Kochi was concerned, he was on my sister-in-law's property. Kirk had closed his property from research, and he got out of Bigfoot research altogether. It was about that time that I went to my sister-in-law's. Beverly Goins is her name. She told me that she had had a sighting. A grey Sasquatch had stood right out in the open leaning on the horse fence, and as she looked out the window, the grey giant simply turned its head and looked right at her, letting her admire him, it seemed, then walked off into the forest. I got the message from Kochi that this was the father of the grey juvenile that had come to him and had helped the deaf and crippled

Bigfoot to hunt and survive. He became one of Kochi's elders and confidants.

But let's return to Kochi's story here while we can. Many things happened in the meantime. I feel that Kochi is helping me write this book as I look back on things when they happened.

Things were happening in the human world that guided us towards moving and leaving the area. I will not get into the negative things that the humans around us did to make that decision, but to say the least, Violet and I had no choice but to move in order to set our lives straight with the world around us, to save our marriage and the way we live, and also to get rid of the humans affecting us in so much of a negative world. Kochi played a part in us escaping from them. Kirk had closed his property up tighter than a fortress. I made one last trip to the farm and walked out to the forest by myself. I felt the love coming from the surrounding forest, and with a heavy heart and with tears running from my eyes, I shouted out into a seemingly empty forest, "I love you guys, and I have to leave. It breaks my heart to have to go, but I must. We will be back, but I do not know when. Bye for now." With that, I turned and walked out of the forest, and sat on the picnic table till my heart calmed and the overwhelming love from all the forest people came to me and helped dry my tears. Kochi came into my mind, and told me, in no certain terms, "I do not like these humans that have destroyed you. I will be watching you and listening in on what is going on, and help if I can. Travel safe, my son, and go with God." That was the last time I stood on that property, as well as any other researcher. We packed everything up, gave up a 78 thousand dollar home, which ruined my credit, and left for Tennessee. We made six trips with truck and cars and trailers back and forth to Tennessee. We ended up in Rockwood.

It was during this time that Kochi came through my mind and told me to trust this forest clan leader, but not to cross him. He was a human guardian as well and would kill to protect the person he guarded. It shook me up some, but I knew we had to finish the journey that was given to me. Kochi let me know that when things got rough he would send a human to help. We just did not know who it was at the time. The other clan leader was called Fox by the human woman. His real name was Haukki (meaning Blocks the Sun). We will call him Fox in the book because that was what he was known as in the human world.

We lived in Rockwood for a year with these people, and things really got bad for us. During that time, my wife had a heart attack from the stress, and I went to the hospital 125 miles away in Nashville. On the way home, I took a detour and stopped at Will Pickens Recreation Area. I got out and walked silently on a trail, climbed up on a boulder, and sat there crying my eyes out; my heart was broken. I felt the energy around me start to change. Kochi came into my mind and calmed me, his voice far away. He said, "The time is at hand. You will meet a woman I am sending to help you." Then he left my mind. That left me confused even more. Then the energy changed again, and everything went silent.

Kochi showed me the forest people that had surrounded me there near the boulder, and that scared the hell out of me. But things changed again, like a fresh rain on the mountain top, and the smell came with it. There was no rain, but the feeling was there. I saw the forest people trying to comfort me, and I knew I had made new friends there as well.

While we were in Tennessee, we fed the Fox clan and saw some of them a few times, but hardly ever had communication from them.

My wife saw them looking in the window in the bathroom while she was showering.

CHAPTER SIX

CHANGE OF SCENERY

In the following months, I started to have a conversation **on the computer** with a woman from Iowa. One day she showed up at the house and saw how we were being treated. She was outraged. In the following days, she asked me if we would move to Iowa. I jumped at the chance to get rid of my problem. I mean, Kochi had told me to go. If a woman tried to help, what else could I do? I walked into our tiny bedroom and told Violet, "We are moving." She was shocked and excited at the same time. It was out of the blue, and I wanted to get her away for health reasons as well as to save my marriage. So the message I had received from Kochi had once again come true. It was amazing. But before we moved, we did research on the Will Pickens area. I had walked into a clan that was there, had waved, took a snapshot of them and walked quickly away. Violet had waved at a small juvenile, and it waved back.

This woman from Iowa sent her son to our home, and I had to leave my six loads of personal furniture, my tools, and my truck. I gave all that away to the lady we had so much trouble with. "Here it is woman; anything to get out of this trouble." It took 15 hours of straight driving to get to Iowa, and at the location there, we had instant contact with the forest people.

I have changed the name of this person to Bonnie, as she does not want to be known, and we respect that. Once we got there, we lived with these folks for 6 weeks before we got our own place 15 miles away. We still visited them off and on after we moved. We respected the rules she gave us. There were many times when we drove out on a deserted gravel road in the middle of the dark forest, set up chairs under a full moon, and watched large shadows cross the road. The heavy walking at night was scary, but at the same time, it was like sitting

around an unlit campfire. Imagine if you will, darkness so full you cannot see two inches in front of your face, and the noise coming from the surrounding forest of animals on the move looking for prey. Then the moon slips out from behind a cloud, and the light from it mixes with the dark branches from the creaking trees above you forming ghostly hands reaching out to you. All of that is broken up by human shapes walking among the light-filtered trees and by the thumping from the feet of the night creatures making its way to your ears while you're sitting there in the darkness waiting to be grabbed. It becomes a chilling thought at times, but humans have the fear factor in them that sets us apart from the rest of the world. We did this many times in Iowa.

When we lived 15 miles from Bonnie's area, we met others in our little village that had the forest people right outside of town, and we got to know that group as well. Kochi had taken a trip out west of Indiana and had come close to going across the mighty Mississippi River to visit us. He came as far as western Illinois but had to turn back towards Indiana because of clan fighting. One of his most important rules was no fighting, and he had to stop it.

It was not long before we had enough money to go back home towards Indiana, and we moved to Greencastle, Indiana.

PART TWO

CHAPTER ONE

MEANDERING...THINGS THAT BELONG

Kochi's thoughts and teachings will be given in this part of the book. I was told not to reveal anything that was said until the right time, and that time has come. After Judy had been robbed of her computer and all her disks and evidence, Kochi felt it was the right time to teach Violet and me the way of things. When Judy mysteriously moved away from the area, Kochi felt even more compelled to inform us of the true way of the forest people.

After returning to Indiana, I went out to Kochi's old area and found Kirk's place locked away for good from research and even neighbors. Judy's home was sitting empty until someone bought it, and the place changed. The fence was down and the trees in the ravine behind the home had been cut down. The old bean field, where the fight had been taken place, was now a corn field.

With great sadness, I went to where I had called the female out on the deserted road and sat in the car. I felt a feeling of doom all over the area. A new clan had moved into Kochi's old area, and they were not friendly to humans. As I was sitting in the car, a feeling of great energy came over me from Kochi. He told me he had moved to a better feeding area because the clan that had belonged to the red rogue wanted to converge with his clan. Kochi had to find a better area for all of them because their size had doubled. While sitting there, I was told to come to the same area and sit and listen to Kochi as he explained the reasons and the rules of the clan and the life he lived.

The next few words will be Kochi's words, not mine. Imagine, if you will, something standing 12 feet tall, all covered with man-hair, with the strength of 15 men and mind-talking to a small human sitting in a car staring off into space, about the clan's lifestyle.

Kochi told me in the lengthy words coming up about the security of the clan and why it was so important. here are his words to the best of my memory:

KOCHI--- "Humans destroy anything they touch in this world. I hate to say this to you, but the truth is the only way we live. From the beginning of time, we were many in this world. Humans lived in the caves and were short men and women that were ugly to the eye. They did not have a way to talk to us and had not the gift of mind-speak, but we did and still do. These cave people, the ones you call Neanderthal, were destructive to each other, hate mongers and brutal in their way of life. They had no concern for each other and those things around them. Many died off. Their diseases killed many of our clan. Some were mate-able and did mate with our younger ones, forming a mighty tribe of different hairy tribes that continue on till today. Those that were purebloods left for the wild and stayed there, bringing two different breeds to the earth. But the wilds, as they called, died out. They did not have the thinking abilities of the tamed. When those died out, the new breed took over and are here today. They saw the way humans evolved and the way they killed to gain things that were important to them. We got more scarce and took to the wilds ourselves in order to survive. We did not want the humans around us, so it became a life and death survival rule that we hid from everything in order to survive. That rule continues today."

I left the area that day with a new perspective and wondered how I was going to make people believe this. Then I realized that what I had heard was something that would take time to write down and leave it there until something told me it was time. I knew this was just the beginning for me, and I knew that with the gift that Kochi had left in my head, I would be able to store all this information and sort out the truth later. Several days later, we took a midnight drive to the deserted road, parked, and rolled down the windows. We just listened to the blackness around us. The feeling of doom was still all around the area. Violet got the feeling that a rogue was around the area that hated humans. We left in a hurry and drove to the old cemetery along the darkened roadway. We parked in the old hilly driveway (this was our gifting area) and turned the car off. We just sat in the darkness soaking the sounds up around us when Kochi entered both of our minds and told us this:

Kochi—"Being that you are here to listen, then listen to me. The reason we still flee humans is because of the way your wars destroy the homes of the 4 legged and the two legged, and your police kill at random. Humans destroy what they can't understand, and their fear takes over. That puts our kind in harm's way since your God has been abounded to almost nothing in your eyes. He has turned his back and let the prophecy begin. God's spirit has fled leaving death and destruction everywhere in the world today, and the modern disease humans spread today, we have no way to cure. We have many breeds who are overcome with the sickness of your sexual lives, and we flee further. Death stalks this forest right now. That is why the doom feeling is upon you. Now you see why we flee once again in order to survive.

The survival tactics we had to learn from the ancients many years ago, we have to use today to stay hidden, but there are times when we come across someone with a pure heart who is brave enough to face us without thinking we are monsters. Pure hearts are like you and Violet who have no motive to harm in any way. You display great respect, and that is why you both were chosen to know the truth."

Hearing this in the stillness of the night, sitting in an old forest graveyard where souls are forgotten and forlorn, it was eerie to say the least. The feeling of doom was getting stronger, and we decided to leave for the night. We drove the 15 miles back home.
 During the next few days we settled in with our daily chores. Sitting at times or taking naps seemed like our routine as we were getting older, but the difference was the dreams that came with them, many dreams, different dreams. There are a few I can talk about because to me it was a learning experience. One was when Kochi took me through the swirl of tunnels, the way of his mind, and the way he sees things, a mindset if you will. Two energies come together to form one bondship. The energy feels like heat from an oven until something is cooked, yet it never burns you. Then it settles at the base of your brain.

During my first trip through one of these wormholes, I was scared to death. Kochi held my mind together. I thought I was losing my mind at first. The speed was fantastic and the colors swirling by were mixed so much that it looked as if I lived in a cameo world of color. It was different at night through the wormhole, as I will call it. At night it was darker, with the darkest black you can imagine, but during the daytime, the brightness of the sun through the maze of colors was amazing. At the end when I stopped and stepped out of the wormhole, I was in a different world, there were many different things that happened in this myriad world.

So here are Kochi's words on this, and what he taught me:

Kochi— "This is the way we go through portals. The energy of our minds taps into the mother earth, the thing you call gravitational latitudes, or EMF. We use all the energies from life forms around us, including the stone around us as a battery to store that energy until it builds up enough energy to tap into the Mother earth. It then pulls us physically through a portal we create with our mind, and we can go to several places that our minds create. The places we go are real. There are several things we use these places for and several reasons we can do what you humans call 'cloaking'. We can hide from humans; bring food back in the winter time to the clan. There are 4-legged and two-legged here as well, but most of them are dangerous. And yes, they can follow us back to the present. That is why you see creatures that you have never seen before in your world. We try to kill them out when they come to this side. Most important of all, we bury a lot of our kind in these other worlds so humans would never find the bones in the graves. And yes, we bury our kind here as well. I will come and take you through many worlds so you see the differences of each of them. For now, wake up and learn."

The first world that Kochi took me through was in the daylight. After stepping out of the portal, I felt sick to my stomach and wanted to throw up, but Kochi prevented me from doing so. The different plants and the brightness of the trees all seemed to give off a living glow of life, much different than our plane of existence. It was as if the plants and trees were trying to physically talk to me.

Then I saw that everything here was doubled, like two suns. The trees next to each other were exact copies of each other, but then only two of each and came another different tree and its brother next to it and so on, a world of doubles. It was a weird feeling. The air I breathed had a funny taste and color to it. I could actually see the air around me. It was a little moist and had a tinge of very light blue to it, like a living organism. Energy was crackling all around me, like eating the cereal that went snap, crackle, and pop. And the animals were beautiful, wild and free, and unafraid. Kochi interrupted the journey and pulled me back through the worm hole, only to tell me, I must go,

++++++++++++++++++++++++++++++++++++
++++++++++++++++++++++++++++++++++

Part two. Chapter two, it is destiny,

++++++++++++++++++++++++++++++++++++

All things must end, and have a start, but yet the end never comes for the soul, just a new beginning, and continues to grow and better itself, the concept of god is indeed real, but a god you cannot see, but again very real, not gods, but one god conceived in the father, the son and holy spirit, died for sins, became back from the dead and moved on with the father, one god with three elements, but yet one, destiny is real as well, it is all written, but a writing we cannot see, but appointed to all humans from above, good or evil, you have the choice in your heart. Destiny is real and whether it is changed on earth in one way or the other, it is still destiny, and will happen, the forest people know this as a fact, and live with it as life without change.

This is a small book, but to me packed with the best stuff I have ever written, all is truths written from Kochi, and my memory of it, so go forth human, and put these truths in your heart and mind and love god and live your life in a positive way, learn from your mistakes, and put them in to your daily life for others to see, and shine, and remember, our here only for a short time, it is destiny.

This book is collected

From several researchers as well from all over the United States, and being already reported as a open public domain, all copyrights for learning purposes can be used to educate the public domain,

Steve Aney